United States Nuclear Regulatory Commission

Protecting People and the Environment

Performance and Accountability Report

FY 2007

MISSION

License and regulate the Nation's civilian use of byproduct, source,

and special nuclear materials to ensure adequate protection of

public health and safety, promote the common defense and security,

and protect the environment.

VISION

Excellence in regulating the safe and secure use and management

of radioactive materials for the public good.

Commissioner Gregory B. Jaczko, Chairman Dale E. Klein, and Commissioner Peter B. Lyons

I am pleased to present the Nuclear Regulatory Commission's *Performance and Accountability Report* for FY 2007. The NRC has again achieved its safety and security performance goals, and continues to position its resources and infrastructure to maintain its strong oversight of existing facilities and to review applications for new nuclear power reactors, license renewals for existing facilities, and a potential high-level waste repository.

The NRC is committed to ensuring that our resources are well managed. This report provides information that demonstrates that NRC's financial and performance data are reliable and complete and the prudent management of the funds entrusted to it by the American public. The auditors have rendered an unqualified opinion on the agency's FY 2007 financial statements. This year, the NRC implemented a number of internal control improvements, eliminated a long-standing material weakness relating to the fee billing process, and evaluated its internal controls, including those relating to financial reporting, and the agency's financial management systems as required by the Federal Managers Financial Integrity Act (FMFIA). There is reasonable assurance that the NRC is in compliance with the FMFIA, with the exception of one material weakness related to implementation of the Federal Information Management Security Act associated with the agency's overall Information Technology (IT) security. The NRC has developed a corrective action plan and will continue to work to eliminate the material control weakness associated with IT security. (See Chapter 1, *Audit Results* and *Management Assurances*). In support of the President's Management Agenda, the NRC is currently cross servicing its Human Resources, Payroll, e-Travel, and Accounting systems. The agency is also in the process of integrating and modernizing its financial systems to enhance further controls, reporting, and decision-making.

The NRC conducts its regulatory responsibilities to enable the use and management of radioactive materials and nuclear fuel for beneficial civilian purposes in a manner that protects public health and safety and the environment, and promotes the security of the Nation. The Commission is proud of this year's performance in achieving the agency's safety and security goals and looks forward to continuing its high-quality service to the American public in FY 2008 and beyond.

Dale E. Klein
November 15, 2007

Yucca Mountain Tunnel Entrance

Management's Discussion and Analysis

NRC Headquarters in Rockville, MD

Tour at Davis Besse Nuclear Power Plant near Oak Harbor, OH

INTRODUCTION

This Performance and Accountability Report represents the culmination of the U.S. Nuclear Regulatory Commission's (NRC) program and financial management processes. It began with mission and program planning, continued through the formulation and justification of NRC's budget to the President and the Congress, through budget execution, and ended with this report on the agency's program performance and use of the resources with which it is entrusted. This report was prepared pursuant to the requirements of the Chief Financial Officers Act, as amended by the Reports Consolidation Act, and covers activities from October 1, 2006, to September 30, 2007.

The NRC places a high importance on keeping the public informed of its activities. Visit our Web site at http://www.nrc.gov to access this report and to learn more about who we are and what we do to serve the American public.

Chapter 1; *Management's Discussion and Analysis*, provides an overview of the NRC. It consists of seven sections—*About the NRC* describes the agency's mission, organizational structure, and regulatory responsibility; the *Program Performance Overview* summarizes the agency's success in achieving its strategic goals, which are further described in Chapter 2; the *Program Performance Results* show the agency's program performance results; *Future Challenges* includes forward-looking information; the *President's Management Agenda* describes the agency progress in "Getting to Green" for five management initiatives; *Financial Performance Overview* highlights the NRC's financial position and audit results contained in Chapter 3; and *Systems, Controls, and Legal Compliance* describes the agency's compliance with key legal and regulatory requirements.

ABOUT THE NRC

The NRC was established on January 19, 1975, as an independent Federal agency regulating commercial and institutional uses of nuclear materials. The Atomic Energy Act, as amended, and the Energy Reorganization Act, as amended, define the NRC's purpose. These acts provide the foundation for the NRC's mission to regulate the Nation's civilian use of byproduct, source, and special nuclear materials to ensure adequate protection of public health and safety, to promote the common defense and security, and to protect the environment.

To fulfill its responsibility to protect public health and safety, the NRC performs three principal regulatory functions. The agency (1) establishes standards and regulations, (2) issues licenses for nuclear facilities and users of nuclear materials, and (3) inspects facilities and users of nuclear materials to ensure compliance with regulatory requirements. These regulatory functions relate to civilian nuclear power plants, other nuclear facilities, and uses of nuclear materials. These include nuclear medicine programs at hospitals; academic activities at educational institutions; research work; industrial applications, such as gauges and testing equipment; and the transport, storage, and disposal of nuclear materials and wastes.

Organization

The NRC is headed by a Commission composed of five members, with one member designated by the President to serve as Chairman. The President appoints each member, with the advice and consent of the Senate, to serve a 5-year term. The Chairman is the principal executive officer and official spokesman for the Commission. The Executive Director for Operations carries out program policies and decisions made by the Commission.

NRC BUDGETARY AUTHORITY, FY 2002-2007

(Dollars in Millions)   **Figure 1**

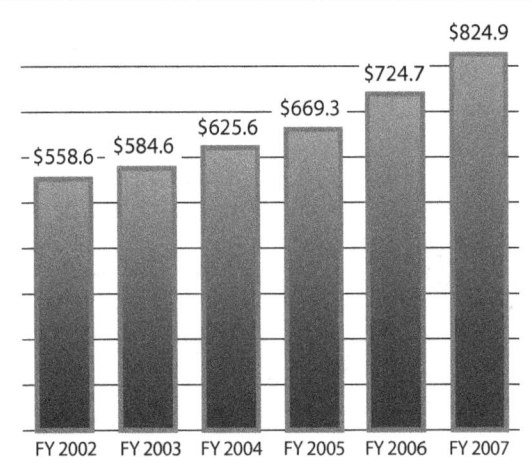

NRC PERSONNEL CEILING, FY 2002-2007

(Staff)   **Figure 2**

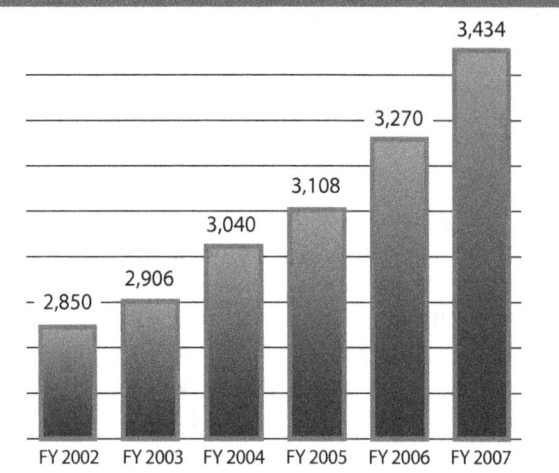

The NRC's headquarters is located in Rockville, Maryland. Four regional offices are located in King of Prussia, Pennsylvania; Atlanta, Georgia; Lisle, Illinois; and Arlington, Texas. The NRC's technical training center is located in Chattanooga, Tennessee. The NRC also has at least two resident inspectors at each of the Nation's nuclear power reactor sites. The NRC's Operations Center which is located in the Headquarters building in Rockville, Maryland is the focal point for the agency's communications with its licensees, State agencies, and other Federal agencies concerning operating events in the commercial nuclear sector. The NRC operations officers staff the Operations Center 24 hours a day. Appendix G to this report presents the NRC organization chart.

The NRC's budget for fiscal year (FY) 2007 was $824.9 million (see Figure 1) with 3,434 full-time equivalent staff (see Figure 2). The NRC recovers most of its appropriations from fees paid by NRC licensees.

The Nuclear Industry

The NRC regulates all activities involved in the commercial use of radioactive materials. From nuclear fuel facilities, which produce the radioactive fuel used in the Nation's 104 nuclear power plants

and other users of nuclear materials, through the safe transportation and disposal of nuclear waste, the NRC's regulatory programs ensure that radioactive materials are used safely and securely throughout this nuclear material cycle. Under the NRC's Agreement States program, 34 states have assumed the majority of regulatory responsibilities for overseeing the activities of industrial, medical and other smaller users of nuclear materials in their states. The NRC works closely with these states to ensure that public safety is maintained. The NRC has a defined set of regulatory practices, knowledge and expertise specific to each activity in the nuclear material cycle to address safety and security issues.

Approximately 20 percent of the Nation's electricity is generated by the 104 NRC-licensed commercial nuclear reactors operating in 31 States (see Figure 3). Since 1994, nuclear electric generation has increased by approximately 22 percent. The NRC oversees 4,369 licenses for medical, academic, industrial and general uses of nuclear materials (see Figure 4). The agency conducts approximately 1,500 health and safety inspections of its nuclear materials licensees annually. In addition, the 34 Agreement States oversee 17,807 licenses. These Agreement States have assumed the majority of regulatory responsibilities for overseeing

U.S. COMMERCIAL NUCLEAR POWER REACTORS

Figure 3

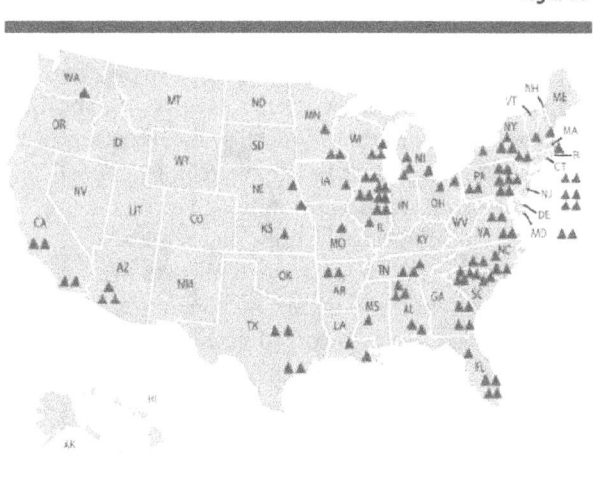

▲ Licensed to Operate (104)

U.S. MATERIALS LICENSEES

□ NRC ■ Agreement States **Figure 4**

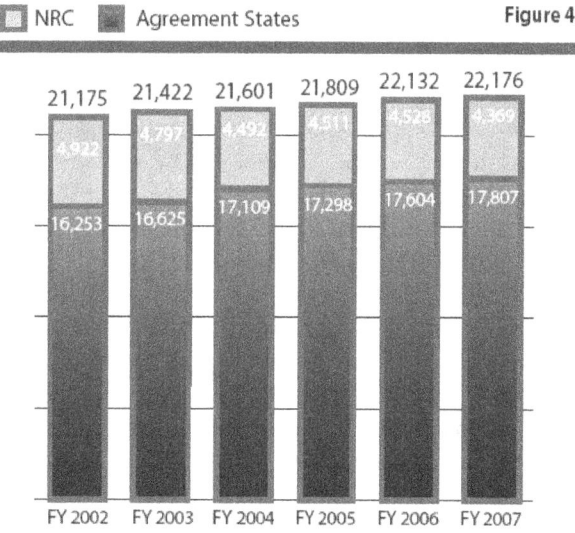

FY 2002 FY 2003 FY 2004 FY 2005 FY 2006 FY 2007

the activities of industrial, medical, and other small users of nuclear material within their borders. The NRC, Agreement States, and their licensees share a common responsibility to protect public health and safety.

Fuel Facilities

Nuclear fuel is derived from milled uranium ore extracted from the earth at uranium mines to produce uranium concentrate called "yellow cake." The yellow cake is converted into uranium hexafluoride gas at a special facility and loaded into cylinders. The cylinders are sent to a gaseous diffusion plant, where uranium is enriched for use as reactor fuel. The enriched uranium is then converted into oxide powder, fabricated into fuel pellets (each about the size of a fingertip),which are loaded into metal fuel rods about 12 feet long and bundled into reactor fuel assemblies at a fuel fabrication facility. Assemblies are then transported to nuclear power plants, non-power research reactor facilities, and naval propulsion reactors for use as fuel. Eight major fuel fabrication and production facilities and two enrichment facilities are licensed to operate in the United States. Because they handle extremely hazardous material, these facilities take special precautions to prevent theft,

diversion by terrorists, and dangerous exposures to workers and the public from this nuclear material.

Reactors

Power plants change one form of energy into another. Electrical generating plants convert heat, the energy of wind or falling water, or solar energy into electricity. A nuclear power plant converts heat into electricity. Other types of heat-conversion plants burn coal, oil, or gas for a heat source that is used to produce electricity. Nuclear energy as it is used in a nuclear power plant cannot be seen. There is no burning of fuel in the usual sense. Rather, energy is given off by the nuclear fuel as certain types of atoms split into pieces. This energy is in the form of fast-moving particles and invisible radiation. As the particles and radiation move through the fuel and surrounding water, the energy is converted into heat. The heat is the useful energy resulting from the splitting of atoms. The radiation energy itself can be hazardous and requires special precautions to protect people and the environment.

Because the fission reaction produces radioactive materials, which can be hazardous, nuclear power plants are equipped with safety systems to protect workers, the public, and the environment. Radioactive materials require careful use because they produce radiation, a form of energy that can damage human cells, and depending on the amount and duration of the exposure, can potentially cause cancer over long periods of time. In a nuclear reactor, most hazardous radioactive substances, called fission byproducts, are trapped in the fuel pellets themselves or in the sealed metal tubes holding the fuel. Small amounts of these radioactive fission byproducts, principally gases, however, become mixed with the water passing through the reactor. Other impurities in the water are also made radioactive as they pass through the reactor. The water is processed and filtered to remove these radioactive impurities and then returned to the reactor cooling system.

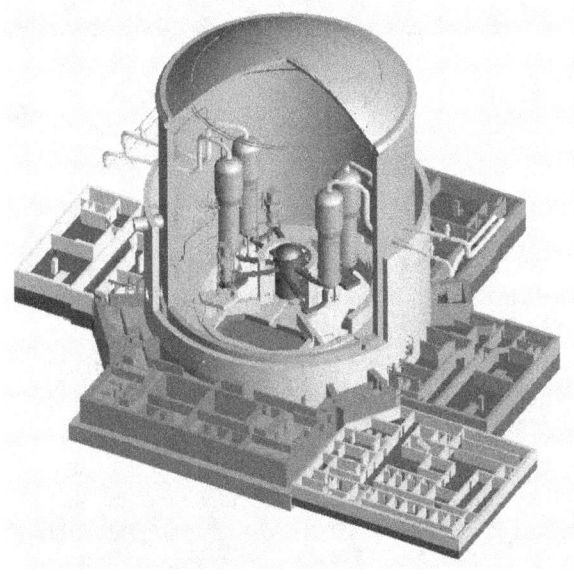

Schematic of a nuclear power reactor

Materials Users

Nuclear materials are used extensively in the medical, academic, and industrial fields. For example, about one-third of all patients admitted to U.S. hospitals are diagnosed or treated using radioisotopes. Most major hospitals have specific departments dedicated to radiation medicine. In all, about 112 million nuclear medicine or radiation therapy procedures are performed annually, with the vast majority used in diagnoses. Radioactive materials used as a diagnostic tool can identify the status of a disease and minimize the need for surgery, reducing the risks from post-operative infection. Radioisotopes give doctors the ability to "look" inside the body and observe soft tissues and organs, in a manner similar to the way X-rays provide images of bones. Radioisotopes carried in the blood also allow doctors to detect clogged arteries or check the functioning of the circulatory system.

The same property that makes radiation hazardous can also make it useful in helping the body heal. When living tissue is exposed to high levels of radiation, cells can be destroyed or damaged so they can neither reproduce nor continue their normal functions. For this reason radioisotopes are used in the treatment of cancer (which amounts to uncontrolled cell division). Although some healthy tissue surrounding a tumor may be damaged during the treatment, mostly cancerous tissue can be targeted for destruction.

Many of today's industrial processes also use nuclear materials. High-tech methods that ensure the quality of manufactured products often rely on radiation generated by radioisotopes. To determine whether a well drilled deep into the ground has the potential for producing oil, geologists use nuclear well-logging, a technique that employs radiation from a radioisotope inside the well to detect the presence of different materials. Radioisotopes are also used to sterilize instruments; to find flaws in critical steel parts and welds that go into automobiles and modern buildings; to authenticate valuable works of art; and to solve crimes by spotting trace elements of poison, among other uses. Radioisotopes can also eliminate dust from film and compact discs as well as static electricity (which may create a fire hazard) from can labels. In manufacturing, radiation can change the characteristics of materials, often giving them features that are highly desirable. For example, wood and plastic composites treated with gamma radiation are used for some flooring in high-traffic areas of department stores, airports, hotels, and churches, because they resist abrasion and ensure low maintenance.

Waste Disposal

During normal operations, a nuclear power plant generates two types of radioactive wastes: high level waste, which consists of used fuel (usually called spent fuel), and low-level wastes, which include contaminated equipment, filters, maintenance materials, and resins used in purifying water for the reactor cooling system. Other users of radioactive materials, such as those discussed above, also generate low-level wastes.

Each type of waste is handled differently. Typically, the spent fuel from nuclear power plants is stored in water-filled pools at each reactor site and at one storage facility in Illinois. However, several utilities have begun using dry cask storage pending final disposal. In this way, spent fuel is stored in heavy metal or concrete containers placed on concrete pads adjacent to the reactor facility. Spent fuel is highly radioactive because it contains the fission byproducts that were created while the reactor was operating. Special procedures are needed in the handling of the spent fuel, since the radiation levels can be very dangerous without proper shielding. The water in the spent fuel storage pool provides cooling and adequately shields workers from the radiation to protect workers in a nuclear plant. Concrete and steel in dry casks provide adequate protection.

Currently most of the spent fuel remains stored at individual plants. Permanent disposal of spent fuel requires a disposal facility that can provide reasonable assurance that the waste will remain isolated for thousands of years. The Department of Energy is developing plans for a permanent disposal facility at Yucca Mountain, Nevada, for spent fuel from nuclear power plants.

PROGRAM PERFORMANCE OVERVIEW

The NRC is developing a new Strategic Plan for FY 2008-FY 2013 that determines the agency's long-term strategic direction. The Commission has approved the framework for the draft Strategic Plan. The Performance and Accountability Report reflects the new goal structure proposed in the agency's draft Strategic Plan and reports performance in support of the Safety and Security strategic goals, as well as Openness, Effectiveness and Management which are referred to as operational goals in this report.

To achieve its goals, the agency is organized into two major programs: Nuclear Reactor Safety, and Nuclear Materials and Waste Safety.

Nuclear Reactor Safety Program

The Nuclear Reactor Safety Program encompasses all NRC efforts to ensure that civilian nuclear power reactor facilities, and research and test reactors are licensed and operated in a manner that adequately protects the public health and safety, and the environment and protects against radiological sabotage and theft or diversion of special nuclear materials. The Nuclear Reactor Safety Program accounted for 74 percent of the agency's costs in FY 2007.

NRC Goals and Programs

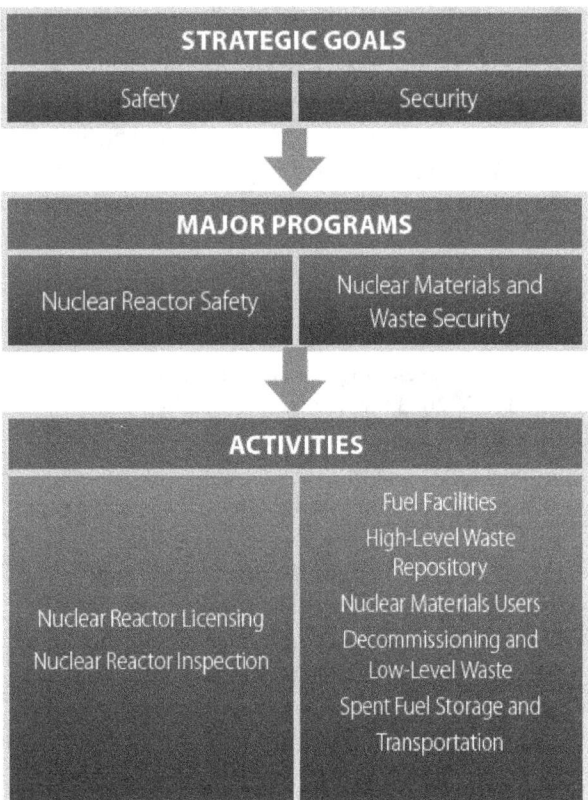

Nuclear Materials and Waste Safety Program

The Nuclear Materials and Waste Safety Program focuses on the safe and secure use of remaining radioactive materials. The Nuclear Materials and Waste Safety Program regulates fuel facilities, medical and industrial nuclear materials users, the disposal of both high-level and low-level waste, the decommissioning of power plants, and the storage and transportation of spent nuclear fuel. The Nuclear Materials and Waste Safety Program accounted for the remaining 26 percent of the agency's costs in FY 2007.

PROGRAM PERFORMANCE RESULTS

STRATEGIC GOAL 1: SAFETY

Ensure protection of public health and safety and the environment.

• Safety: Ensure protection of public health and safety and the environment

Safety is the primary goal of the NRC. The agency achieves its safety goal by ensuring that the performance of licensees is at or above acceptable safety levels. NRC programs work in conjunction with our licensees in a partnership to achieve the safety goal. The NRC licensees are responsible for designing, constructing, and operating nuclear facilities safely, while regulatory oversight of the licensees is the responsibility of the NRC. The strategic outcomes, below, are specific hazards that NRC activities are designed to mitigate against.

Strategic Outcomes:

• No nuclear reactor accidents.

• No inadvertent criticality events.

• No acute radiation exposures resulting in fatalities.

• No releases of radioactive materials that result in significant radiation exposures.

• No releases of radioactive materials that cause significant adverse environmental impacts.

NRC PERFORMANCE MEASURE RESULTS

FY 2007 Safety Goal

Performance Measures	2002	2003	2004	2005	2006	2007
1. Number of new conditions evaluated as red by the Reactor Oversight Process is ≤3.	2	1	1	0	0	0
2. Number of significant accident sequence precursors of a nuclear reactor accident is zero.	1	0	0	0	0	0
3. Number of operating reactors with integrated performance that entered the Manual Chapter 0350 process, or the multiple/repetitive degraded cornerstone column or the unacceptable performance column of the Reactor Oversight Program Action Matrix, with no performance exceeding Abnormal Occurrence Criterion I.D.4 is ≤4.	3	2	1	0	0	1
4. Number of significant adverse trends in industry safety performance with no trend exceeding the Abnormal Occurrence Criterion I.D.4 is ≤1.	0	0	0	0	0	0
5. Number of events with radiation exposures to the public and occupational workers that exceed Abnormal Occurrence Criterion I.A is:						
Reactors: 0	0	0	0	0	0	0
Materials: ≤3	0	0	0	1	0	0
Waste: 0	0	0	0	0	0	0
6. Number of radiological releases to the environment that exceed applicable regulatory limits is:						
Reactors: 3	0	0	0	0	0	0
Materials: 2	4	0	1	0	0	0
Waste: 0	0	0	0	0	0	0

FY 2007 Security Goal

Performance Measures	2002	2003	2004	2005	2006	2007
1. Unrecovered losses or thefts of risk-significant radioactive sources is zero.	0	0	0	0	0	0
2. Number of substantiated cases of theft or diversion of licensed, risk-significant radioactive sources or formula quantities of special nuclear material; or attacks that result in radiological sabotage is zero.	0	0	0	0	0	0
3. Number of substantiated losses of formula quantities of special nuclear material or substantiated inventory discrepancies of formula quantities of special nuclear material that are caused by theft or diversion or by substantial breakdown of the accountability system sabotage is zero.	0	0	0	0	0	0
4. Number of substantial breakdowns of physical security or material control that significantly weaken the protection against theft, diversion, or sabotage is less than one security events and incidents that exceed Abnormal Occurrence Criteria I.C.2B4 is ≤4.	0	0	0	0	0	0
5. Number of significant unauthorized disclosures of classified and/or safeguards information is zero.	0	0	0	0	0	0

FY 2007 Operational Goals and Associated Performance Measures

Measure	2002	2003	2004	2005	2006	2007
Goal 3: Openness						
1. 90% of surveyed stakeholders that perceive the NRC to be open in its processes.		New measure in FY 2006			N/A	94%
2. 88% of selected openness output measures that achieve performance targets. **Not Achieved**		New measure in FY 2006			50%	**66%**
a. Ninety percent of stakeholder formal requests for information receive an NRC response within 60 days of receipt.		New measure in FY 2006			100%	100%
b. The NRC achieves a 72% user satisfaction score for the agency's public Web site greater than or equal to the Federal Agency Mean score based on results of the yearly American Customer Satisfaction Index for Federal Web sites. **Not Achieved**		New measure in FY 2006			70%	**71%**

FY 2007 Operational Goals and Associated Performance Measures (continued)

Measure	2002	2003	2004	2005	2006	2007
c. Complete 50% of Freedom of Information Act requests in 20 days (median).	New measure in FY 2006				61%	67%
d. Issue 90% of Director's Decisions under 2.206 within 120 days.	New measure in FY 2006				100%	100%
e. Make 90% of Final Significance Determination Process Determinations within 90 days for all potentially greater than green findings:	New measure in FY 2006				92%	100%
f. 90% of stakeholders believe they were given sufficient opportunity to ask questions or express their views.	New measure in FY 2006				90%	96%
g. At least 90% of Category 2 and 3 meetings on regulatory issues for which public notices are issued at least 10 days in advance of the meeting	New measure in FY 2006				92%	93%
h. 90% of non-sensitive, unclassified regulatory documents generated by the NRC and sent to the agency's Document Processing Center that are released to the public by the 6th working day after the date of the document. **Not Achieved**	New measure in FY 2006				63%	**75%**
i. 90% of non-sensitive, unclassified regulatory documents received by the NRC that are released to the public by the 6th working day after the document is added to the ADAMS main library. **Not Achieved**	New measure in FY 2006				77%	**87%**
Goal 4: Effectiveness						
1. 70% of selected processes deliver efficiency improvements. **Not Achieved**	New measure in FY 2006				25%	**60%**
a. 10% reduction in the average enforcement processing time for Handling Discrimination Allegations. **Not Achieved**	New measure in FY 2006				N/A	**0%**
b. Eliminate the requirement for license renewal and approve a living license for the two category III facilities which have been renewed in FY 2006 and FY 2007. **Not Achieved**	New measure in FY 2006				Not Elim- inated	Not Elim- inated
c. Improve the timeliness of the review process for nuclear power reactor License Termination Plans by at least 30% over 3 years (FY 2006-FY 2008) as compared to the historical average.	New measure in FY 2006				N/A	N/A
d. Reduce resources expended in support of each interagency exercise by 5% while still accomplishing agency goals for each exercise.	New measure in FY 2006				N/A	5%
e. Implement process enhancements to permit improvement f the reactor rulemaking petition timeliness by 5%.	New measure in FY 2006				N/A	5%
f. Achieve an average 5% reduction on license renewal resources for applications completed in FY 2007.	New measure in FY 2006				N/A	5%
2. No more than one instance per program where licensing or regulatory activities unnecessarily impede the safe and beneficial uses of radioactive materials.	New measure in FY 2006				0	0
Goal 5: Management						
1. 70% of selected support processes deliver efficiency improvements. **Not Achieved**	New measure in FY 2006				50%	**0%**
a. Percent reduction in time (10% in FY 2006 and 5 % in FY 2007) necessary to add or remove employees from drug testing pool.	New measure in FY 2006				10%	N/A
b. 5% reduction of agency FTE used to develop and submit the FY 2008 and FY 2009 performance budgets. **Not Achieved**	New measure in FY 2006				0%	**12%**
c. Issue offer letter 80% of the time within 45 work days of the closing date of the announcement. **Not Achieved**	New measure in FY 2006				67%	**31%**
2. 70% of selected NRC management programs deliver intended outcomes.	New measure in FY 2005			60%	80%	100%
a. Infrastructure management program	New measure in FY 2005			100%	100%	100%
b. Financial Management & Budget and Performance Integration program	New measure in FY 2005			67%	67%	88%
c. Expanded electronic government program.	New measure in FY 2005			50%	75%	75%
d. Management of Human Capital program	New measure in FY 2005			80%	100%	80%
e. Internal Communication program: 100% of activities achieve their targets	New measure in FY 2005			100%	100%	N/A

FY 2007 Results

The NRC achieved all five of its Safety goal strategic outcomes shown above in FY 2007. The NRC also uses six performance measures, to determine whether it has met its Safety goal. All six performance measure targets were met in FY 2007.

Three of the measures focus on performance at individual nuclear power plants. Inspection results show that all of the nuclear power plants are operating safely. However, one measure, *Number of operating reactors with integrated performance that entered the Manual Chapter 0350 process, or the multiple/repetitive degraded cornerstone column or the unacceptable performance column of the Reactor Oversight Program Action Matrix, with no performance exceeding Abnormal Occurrence Criterion I.D.4*, shows an increase from 0 to 1 during FY 2007. One reactor met the conditions in this measure during FY 2007. The Palo Verde Unit 3 entered the multiple/repetitive degraded cornerstone column because of safety system equipment problems and the licensee was not effective at addressing and fixing them. NRC inspections identified the issue and brought it to the attention of licensee management for correction. Palo Verde is scheduled for a significant site review in FY 2008. In addition, another measure uses risk analysis to determine safe operations shows that none of the plants experienced a significant precursor, defined as an event which has a 1 in 1,000 probability of leading to substantial damage to the reactor fuel. This mea-sure indicates that not only were the plants operated safely, but the events that did occur were of relatively minor significance.

The fourth measure tracks the trends of several key indicators of nuclear power plant safety. This mea-sure is the broadest measure of the safety of nuclear power plants, incorporating the performance results from all plants to determine industry average results. The measure results show that there were no statistically significant adverse trends in any of the indicators in FY 2007.

The last two safety measures track harmful radiation exposures to the public and occupational workers, and radiation exposures that harm the environment. None of these measures exceeded their targets in FY 2007.

STRATEGIC GOAL 2: SECURITY

Ensure the secure use and management of radioactive materials.

- Security: Ensure the secure use and management of radioactive materials

The NRC must remain vigilant in ensuring the security of nuclear facilities and materials in an elevated threat environment. The agency achieves its common defense and security goal using licensing and oversight programs similar to those employed in achieving its safety goal.

Strategic Outcome:

- No instances where licensed radioactive materials are used domestically in a manner hostile to the security of the United States.

FY 2007 Results

The NRC achieved its one Security goal strategic outcome shown above in FY 2007. The NRC also uses five Security goal performance measures, in addition to the Security goal strategic outcomes to determine whether we have met our Security goal. All five performance measure targets were met in FY 2007. The first performance measure is whether there were any unrecovered losses or thefts of risk-significant radioactive sources. The measure ensures that those radioactive sources that the agency has determined to be risk-significant to the public health and safety are accounted for at all times. The ability to account for these sources is critical to secure the critical infrastructure of the nation from "dirty bomb" attacks, or other means of radiation dispersal.

The second, third, and fourth performance measures evaluate the number of significant security events and incidents that occur at NRC licensed facilities. These measures determine whether nuclear facilities are maintaining adequate protective forces to prevent theft or diversion of nuclear material or sabotage, whether systems in place at licensee plants are accurately accounting for the type and amount of materials which are processed, utilized, or stored, and whether the facilities are accounting for special nuclear material at all times and that no losses of this material has occurred. There were no events that met the conditions for this measure in FY 2007.

The last security measure is whether there were any significant unauthorized disclosures of classified and/or safeguard information that may cause damage to national security or public safety. This measure determines whether classified information or safeguards information is stored and utilized in such a way as to prevent its disclosure from the public, terrorist organizations, other nations, or personnel without a need to know. Unauthorized disclosures can harm national security or compromise public health and safety. The measure also determines whether controls are in place to maintain and secure the various devices and systems (electronic or paper based) which the agency and its licensees use to store, transmit, and utilize this information. There were no documented disclosures of this type of information during FY 2007.

Operational Goals: Openness, Effectiveness, and Management

Openness

Under Openness, the agency achieved one of two performance measures. The agency achieved its performance measure of stakeholders that perceive the agency to be open in its processes, with a survey score of 94 percent. However, it missed the performance

measure target of 88 percent for selected openness output measures that achieve their output targets. The agency achieved a score on this measure of 66 percent because it missed three output measure targets.

The first missed output measure under the performance measure, *The NRC achieves a 72 percent user satisfaction score for the agency's public Web site greater than or equal to the Federal Agency Mean score based on results of the yearly American Customer Satisfaction Index for Federal Web sites* missed its target by one percentage point with a score of 71 percent. The agency will continue to work on the Web site to improve it so it can meet the target.

The second output measure that missed its target, *90 percent of non-sensitive, unclassified regulatory documents generated by the NRC and sent to the agency's Document Processing Center that are released to the public by the 6th working day after the date of the document,* showed significant improvement over the 2006 result increasing from 63 percent to 75 percent yet still fell below the 90 percent target. The agency will continue to review its internal review processes to find efficiencies so that the amount of time necessary to release documents can be reduced.

The third output measure that missed its target, *90 percent of non-sensitive, unclassified regulatory documents received by the NRC that are released to the public by the 6th working day after the document is added to the ADAMS main library,* also showed significant improvement from 77 percent to 87 percent. However, it was still 3 percent below the 90 percent target. The agency will continue its staff training efforts to close the gap on this measure

Effectiveness

Under Effectiveness, the agency achieved one of two performance measures. The agency missed two of five targets associated with the first measure *70 percent of selected processes deliver efficiency improvement.* The first output under the measure that was missed called for a *10 percent reduction in the average enforcement processing time for handling discrimination allegations.* Two discrimination cases were processed during FY 2007. They took an average of 236 days to process.

The agency was not able to meet the 10 percent reduction in processing time due to the complexity of utilizing alternative dispute resolution in the case. The direct costs associated with post-investigation alternative dispute resolution are greater than the costs for processing traditional enforcement actions. Efficiencies have been made and continue to be made in the alternative dispute resolution process which should allow the agency to reduce the processing time for these cases. The second missed output under the measure was to *Eliminate the requirement for license renewal and approve a living license for the two Category III facilities which have been renewed in FY 2006 and FY 2007.* The agency has not approved a living license for these facilities yet.

The agency met its second effectiveness performance measure regarding the number of instances per program where licensing or regulatory activities unnecessarily impede the safe and beneficial uses of radioactive materials.

Management

Under Management, the agency achieved one of two performance measures. The first measure, to deliver efficiency improvements for selected support processes, was not achieved. Both outputs under the measure were missed. The first target that was missed was to *issue an offer letter to new employees within 45 work days of the closing date of the employment announcement 80 percent of the time.* However, offer letters were issued within 45 days only 31 percent of the time in FY 2007. The NRC undertook a Lean Six

Sigma study during the second quarter of FY 2007 to evaluate the hiring process from the closing date of the announcement to the offer date and develop recommendations to help streamline that process. The agency is currently leading a separate effort to implement the recommendations made by the Lean Six Sigma study workgroup and to develop a plan to assess NRC's progress towards reducing the hiring time frame to meet the 45-day target.

The second target that was missed was a *5 percent reduction of agency Full Time Equivalents (FTE) used to develop and submit the FY 2008 and FY 2009 performance budgets.* The agency has experienced a large growth in FTEs within the last year due to the New Reactor Program ramping up to receive applications from licensees to develop and construct new reactors. As a result, additional budget staff was hired to manage the program which resulted in the agency exceeding the target for this measure. However, the Office of the Chief Financial Officer is currently developing a new budget process as directed by the Commission and it is anticipated that there should be a reduction in FTEs to develop the FY 2010 Performance Budget.

The second Management performance measure assessed the agency's performance in delivering outcomes in four management programs: infrastructure management, financial management, information technology management, and human capital management. These programs were able to meet their intended outcomes based on successfully meeting the sub-measures within each program.

Program Assessment Rating Tool Results

Another important measure of the effectiveness of the agency's programs are Program Assessment Rating Tool (PART) reviews of the agency's program activities conducted by the Office of Management and Budget. The results of the agency's PART scores are shown below:

Program	Year	Score	Rating
Reactor Inspection and Performance Assessment	2003	89	Effective
Fuel Facilities Licensing and Inspection	2003	89	Effective
Nuclear Materials Users Licensing and Inspection	2004	93	Effective
Reactor Licensing	2005	74	Moderately Effective
Spent Fuel Storage and Transportation Licensing and Inspection	2005	89	Effective
Decommissioning and Low-Level Waste	2007	91	Effective
High-Level Waste Repository	2007	87	Effective

Brief discussions of the PART analyses completed in FY 2007 are presented below.

Decommissioning and Low-Level Waste

This program was rated effective in FY 2007. The program earned high scores for Program Purpose and Design and for Program Management. The PART noted that the purpose was clear and the program uses regular independent assessments have helped the program to become more results-focused. The program achieves its long-term safety and security goals with respect to the safe management and cleanup of an increasing number of NRC licensed sites that use radioactive material.

The improvement plan for the program includes developing better linkage of budget requests to the program's success in accomplishing annual and agency long-term goals to make clear how funding affects program accomplishment. Another follow-up action is to improve quantitative measurements of efficiency, including baselines and annual targets to better demonstrate year-to-year performance trends.

High-Level Waste Repository

This program is rated effective in FY 2007. The program earned high scores for Program Purpose and Design and for Program Management. The PART noted that the purpose was clear and the program used regular, independent assessments to help the program become more results focused and satisfying NRC's Nuclear Waste Policy Act responsibilities and pre-licensing functions. The PART also indicated that the program has made significant progress toward meeting the goal of establishing a regulatory system to ensure the repository achieves long-term safety and security goals.

The improvement plan for the program includes developing better linkage of budget requests to the program's success in accomplishing annual and agency long-term goals to make clear how funding affects program accomplishment. Another follow-up action is to improve quantitative measurements of efficiency, including baselines and annual targets to better demonstrate year-to-year performance trends.

FUTURE CHALLENGES

The NRC ensures that the health and safety of the American public and the environment is adequately protected from any harmful effects of using nuclear materials. The industry has experienced a substantial improvement in safety at nuclear power plants over the past twenty years as both the nuclear industry and the NRC have gained substantial experience in the operation and maintenance of nuclear power facilities. Improvements in safety have occurred at a time when nuclear power generation has increased significantly, from 610,000 gigawatt hours in CY 1993 to approximately 787,000 gigawatt hours in CY 2006. However, despite the excellent safety and security record in the industry, the agency cannot rest on its achievements. The primary challenges faced by the agency are the expected large number of new nuclear plants expected to apply for licenses, the safe disposal of high-level nuclear waste, materials degradation, and security at nuclear facilities.

New Nuclear Power Plants

With increased concerns about the continued availability and cost of oil as well as environmental damage caused by coal burning electrical plants, the amount of electricity supplied by nuclear power is likely to increase substantially in the future. The agency expects a large number of applications for the construction of new power plants to be filed over the next few years. The last nuclear power plant construction permit was issued in 1977. The agency's primary challenge is to license the next generation of nuclear reactors to ensure that they will operate safely while providing electricity required by the Nation for economic growth. These new reactor designs require detailed analysis of their vulnerability to accidents and security compromises, as well as development of inspection procedures, tests, analysis, and acceptable criteria for their construction. The NRC is also evaluating commercial gas centrifuge facilities that utilize new methods of enriching nuclear fuel to supply fuel for the reactors.

Safe Disposal of High-Level Waste

The NRC also faces a major challenge as the Department of Energy prepares an application to establish the Nation's first repository for high-level radioactive waste at Yucca Mountain, NV. Safely disposing of the waste from nuclear power plants is vital to protect public health and the environment. Lack of storage options would become a major roadblock for the continued growth of the industry if storage options are not available. The Department of Energy has indicated that a license application may not be filed until mid-2008. The NRC's review of this application will require the evaluation of a wide-range of technical and scientific issues and resolution of many difficult regulatory concerns. Safe and secure interim storage capacity must be ensured until a repository is licensed and ready to receive high-level nuclear waste. In addition to the storage of nuclear waste, safely transporting spent nuclear fuel is a significant issue for the public and the agency. Most nuclear waste is now safely and securely stored at the reactor sites. More than 1,300 spent fuel shipments regulated by the NRC have been safely transported in

the United States in the past 25 years. It is anticipated that the bulk of the nuclear waste now stored at the reactor sites will eventually be moved to a permanent storage site. Therefore, the agency must be able to assure the public that all movements of nuclear waste, including those to a permanent storage site, will be safe and secure.

Security at Nuclear Facilities

In addition to the safety issues, the security of nuclear materials is of paramount importance to the Nation. The agency continues to improve the requirements which better ensure the security of nuclear materials and facilities. The threat faced by the Nation from those seeking to steal classified information has become more urgent in recent years. Nuclear facilities have implemented increased security measures, including force-on-force exercises, to help ensure protection of this vital national infrastructure. Nuclear facilities are among the most secure facilities in the nation. Intelligence is constantly monitored to determine the level of threat faced by nuclear facilities.

PRESIDENT'S MANAGEMENT AGENDA

INITIATIVE 1

Strategic Management of Human Capital

The NRC's ability to accomplish its mission depends on the availability of a highly skilled and experienced workforce. The Commission is proud of the NRC's ranking as the Best Place to Work in the Federal Government based on responses to the 2006 Federal Human Capital survey. To ensure flexibility in its management of human capital and to promote efficiency, the NRC has streamlined recruitment and the review and approval process for relocation and retention incentives. Through the use of an automated strategic workforce planning tool, the NRC is able to determine what critical skill/knowledge gaps exist and can gear its recruitment and other programs (e.g., grants and fellowships) appropriately.

INITIATIVE 2

Budget and Performance Integration

The NRC continues to make progress in achieving budget and performance integration in accordance with the President's Management Agenda. This progress includes developing and ultimately adopting new outcome-based performance measures and revising the agency's Strategic Plan, accurately monitoring program performance, and integrating performance goal information with associated costs.

INITIATIVE 3

Competitive Sourcing

One of the NRC's corporate management strategies is to acquire goods and services in an efficient manner. To achieve this, the NRC established output measures associated with the implementation of the competitive sourcing initiative under the President's Management Agenda, adopted a performance-based approach to contracting, and posted procurement synopses on the agency's Web site. The NRC uploaded its Year 2007 Federal Activities Inventory Reform Act inventory in the Office of Management and Budget's Workforce Inventories Tracking System on June 29, 2007. In accordance with NRC's Competitive Sourcing Plan, potential commercial activities have been identified to be studied to determine which are appropriate for public-private competition. The NRC completed three business case analyses by the end of FY 2007.

INITIATIVE 4

Expanded Electronic Government

The NRC has aligned its information technology investments with the Federal Government's Electronic Government program (e-gov). NRC has completed migration to a number of e-gov services and is in the process of migrating to others. NRC has also institutionalized internal processes to ensure effective use and compliance with e-gov. The NRC emphasizes enterprise architecture in its systems development life cycle methodology and has a Project Management Methodology in place. The Project Management Methodology provides full life cycle guidance for the agency, providing guidance for enterprise architecture, capital planning and investment control (CPIC), infrastructure development, and life cycle management processes. An Information Technology Senior Advisory Council, comprising senior business managers, plays an integral role in ensuring technology investments align to the agency's mission and goals and in establishing priorities.

INITIATIVE 5

Improved Financial Management

The agency's goals for improved financial management include providing reliable, transparent, useful, and timely information to stakeholders and for management decision making; maintaining adequate controls; and implementing integrated and flexible systems to meet the agency's reporting needs. This will ensure that NRC's financial assets are adequately protected consistent with risk. The agency is implementing a new core financial management system hosted by a shared service provider based on Web-enabled commercial off-the-shelf software. The new system will combine the functionality of the existing core accounting, license fee billing, cost accounting, allotment/allowance financial plan, and the capitalized property systems into a single enterprise-wide system. This systems strategy will result in more efficient transaction processing utilizing electronic workflow management, greater access to information through the use of ad-hoc reporting tools, and improved overall system performance. An integrated financial management system will also improve internal controls by eliminating multiple data transfers between stand alone systems and the resultant manual reconciliations currently performed to ensure data integrity.

FINANCIAL PERFORMANCE OVERVIEW

As of September 30, 2007, the financial condition of the NRC was sound with respect to having sufficient funds to meet program needs and adequate control of these funds in place to ensure obligations did not exceed budget authority. The NRC prepared its financial statements in accordance with the accounting standards codified in the Statements of Federal Financial Accounting Standards (SFFAS) and OMB Circular A-136, Financial Reporting Requirements.

Sources of Funds

The NRC has two appropriations, Salaries and Expenses and Office of the Inspector General and funds for both appropriations are available until expended. The NRC's total new FY 2007 budget authority was $824.9 million. Of this amount, $816.5 million was for the Salaries and Expenses appropriation and $8.4 million was for the Office of the Inspector General appropriation. This represents an increase in new budget authority of $83.4 million over FY 2006 ($83.3 million for the Salaries and Expenses appropriation and $0.1 million for the Office of the Inspector General appropriation). In addition, $74.8 million from prior-year appropriations, $3.5 million from prior-year reimbursable work, and $7.7 million for new reimbursable work to be performed for others was available to obligate in FY 2007. The sum of all funds available to obligate for FY 2007 was $910.9 million, which is a $101.9 million increase over the FY 2006 amount of $809.0 million.

The Omnibus Budget Reconciliation Act of 1990 (OBRA-90), as amended, required the NRC to collect fees to offset approximately 90 percent of its new budget authority, less the amount appropriated to the NRC from the Nuclear Waste Fund and amounts appropriated for waste incidental to reprocessing and generic homeland security for FY 2007. The NRC collected $668.8 million in reactor and material fees in FY 2007 (see Figure 5). For FY 2006, OBRA-90 required NRC to collect approximately 90 percent of

its new budget authority, excluding appropriations from the Nuclear Waste Fund and amounts appropriated for waste incidental to reprocessing.

SOURCES OF FUNDS
(In Millions)   **Figure 5**

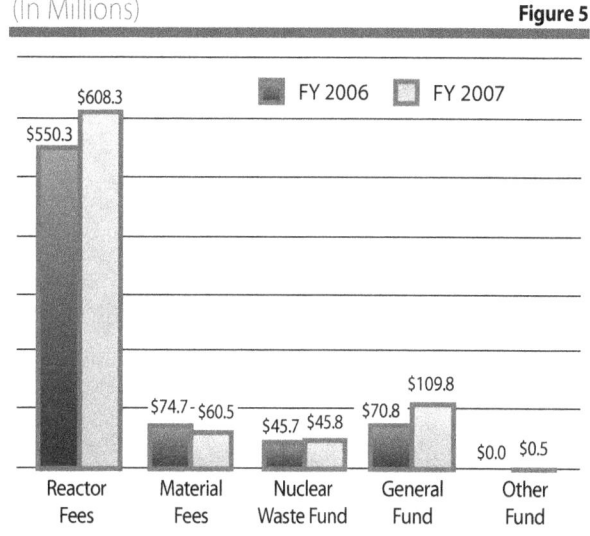

Uses of Funds by Function

The NRC incurred obligations of $838.8 million in FY 2007, which was an increase of $104.0 million over FY 2006. Approximately 56 percent of obligations were used for salaries and benefits. The remaining 44 percent was used to obtain technical assistance for the NRC's principal regulatory programs, to conduct confirmatory safety research, to cover operating expenses, (e.g., building rentals, transportation, printing, security services, supplies, office automation, training), staff travel, and reimbursable work (see Figure 6). The unobligated budget authority available at the end of FY 2007 of $72.2 million, decreased compared to the FY 2006 amount of $74.3 million. Of this $72.2 million, $6.6 million is for reimbursable work and $65.6 million is available to fund critical NRC needs in FY 2008.

Audit Results

The NRC received an unqualified audit opinion on its FY 2007 financial statements. In FY 2007, the auditors identified a continuing material weakness in the agency's information system-wide security

USES OF FUNDS BY FUNCTION

(In Millions)   **Figure 6**

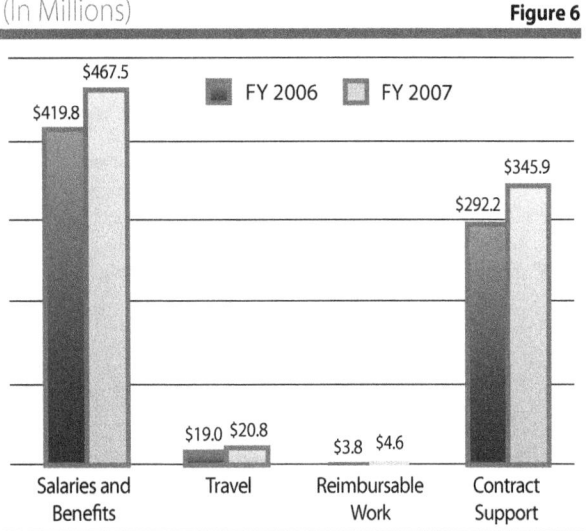

controls related to an independent evaluation of the NRC's implementation of the Federal Information Security Management Act (FISMA). The FISMA report identified two significant deficiencies related to a lack of contingency plan testing for information security systems, and a lack of certification and accreditation for most of the agency's major information systems. These deficiencies were also identified as a material weakness in the agency's Federal Managers' Financial Integrity Act (Integrity Act) assurance statement. The NRC plans to have contingency plan testing completed during FY 2008 and one-half of the systems certified and accredited by September 2008, with the remaining systems being certified and accredited by September 2009.

In FY 2004, FY 2005, and FY 2006, the auditors identified a material weakness concerning the Fee Billing System and the quality assurance process over fee billing. In FY 2007, the auditors downgraded this to a significant deficiency. NRC management has classified the fee billing process as a control deficiency in the annual Integrity Act assurance statement based on the corrective actions to implement compensating controls during the current and prior fiscal years (see Chapter 1, *Management Assurances*). In FY 2006, the Fee Billing System was also identified as a substantial non-compliance with the Federal Financial Management Improvement Act (Improvement Act).

In FY 2007, the Fee Billing System and Human Resources Management System are substantially non-compliant with the Improvement Act due to a lack of current certification and accreditation. In addition, a general support system, which all financial management systems either reside on or rely on, does not have a current certification and accreditation and did not have the annual contingency plan tested. Although there may be a potential risk with security controls, there are a number of existing mitigating controls that provide NRC management reasonable assurance that the financial data resulting from financial management systems is accurate. NRC will continue to improve internal controls by implementing and monitoring corrective actions during the agency's internal control assessment.

The auditors closed the remaining prior-year reportable condition concerning 10 CFR Part 170 hourly rates for license fees. A summary of the Financial Statement Audit Results is included in Appendix D.

Limitations of the Financial Statements

The principal statements have been prepared to report the financial position and results of operations of the NRC, pursuant to the requirements of the Chief Financial Officers Act of 1990, as amended by the Government Management and Reform Act of 1994. These statements have been prepared for the books and records of the NRC in accordance with the formats prescribed by the Office of Management and Budget and in accordance with accounting principles generally accepted in the United States. However, these statements differ from the financial reports used to monitor and control budgetary resources that are prepared from the same books and records. The principal statements should be read with the realization that they are for a sovereign entity, liabilities not covered by budgetary resources cannot be liquidated without enactment of an appropriation, and the payment of all liabilities other than for contracts can be abrogated by the sovereign entity. Other limitations are discussed in the footnotes to the principal statements.

The NRC's FY 2007 financial statements were audited by R. Navarro and Associates, Inc., under contract to the NRC Office of the Inspector General.

Financial Statement Highlights

The NRC's financial statements summarize the financial activity and financial position of the agency. The financial statements, footnotes, and required supplementary information, appear in Chapter 3, *Financial Statements and Auditors' Report*. Analysis of the principal statements follows:

ASSET SUMMARY (IN MILLIONS)

	FY 2007	FY 2006
Fund Balance with Treasury	$ 356.4	$ 281.7
Accounts Receivable, Net	93.9	75.2
Property & Equipment, Net	31.8	26.9
Other	3.3	2.3
Total Assets	$ 485.4	$ 386.1

Analysis of the Balance Sheet

The NRC's assets were approximately $485.4 million as of September 30, 2007. This is an increase of $99.3 million from the end of FY 2006. The assets reported in NRC's Balance Sheet are summarized in the accompanying table.

The Fund Balance with Treasury represents the NRC's largest asset of $356.4 million as of September 30, 2007, an increase of $74.7 million from the FY 2006 year-end balance. This balance accounts for approximately 73 percent of total assets and represents appropriated funds, collected license fees, and other funds maintained at the U.S. Treasury to pay current liabilities. The increase in Fund Balance with the U.S. Treasury is primarily due to an $83.4 million increase in new budget authority offset by a $77.8 million increase in expenditures, a $58.1 million increase in fees collected, and a $16.9 million increase in the fund balance carryover from the prior year.

Accounts Receivable, Net, as of September 30, 2007, was $93.9 million which includes an offsetting allowance for doubtful accounts of $4.7 million. This is a 25 percent increase from the FY 2006 year-end Accounts Receivable, Net, balance of $75.2 million. The increase was primarily due to an increase in annual fees for reactor licensing and an increase in the hourly rates for materials and facilities inspection fees. The value of Property, Plant, and Equipment, Net, was $31.8 million, representing 7 percent of total assets. The majority of this balance represents information technology software and leasehold improvements.

LIABILITIES SUMMARY (IN MILLIONS)

	FY 2007	FY 2006
Accounts Payable	$ 27.7	$ 31.2
Federal Employee Benefits	6.8	7.4
Other Liabilities	169.7	134.9
Total Liabilities	$ 204.2	$ 173.5

The NRC's liabilities were $204.2 million as of September 30, 2007. The accompanying table shows an increase in Total Liabilities of $30.7 million from the FY 2006 year-end balance of $173.5 million. This increase is primarily due to the increase in the liability that relates to future collections, which will be paid to the U.S. Treasury. Other Liabilities include $93.4 million for recoveries from accounts receivable, $38.3 million for accrued annual leave, and $16.0 million for accrued salaries to employees. Of the agency's liabilities, $46.8 million were not covered by budgetary resources, which is a slight increase over the balance as of September 30, 2006. The liabilities not covered by budgetary resources include unfunded accrued annual leave and future workers' compensation.

NET POSITION SUMMARY (IN MILLIONS)

	FY 2007	FY 2006
Unexpended Appropriations	$ 254.0	$ 193.7
Cumulative Results of Operations	27.2	18.9
Total Net Position	$ 281.2	$ 212.6

The difference between Total Assets and Total Liabilities, Net Position, was $281.2 million as of September 30, 2007. This is an increase of $68.6 million from the FY 2006 year-end balance. Net Position is comprised of two sections: Unexpended Appropriations and Cumulative Results of Operations. Unexpended Appropriations is the amount of authority granted by Congress that has not been expended. The increase of Unexpended Appropriations of $60.3 million for FY 2007 is primarily due to funding for the expected added volume of new reactor licensing activities.

NET COST OF OPERATIONS (IN MILLIONS)

	FY 2007	FY 2006
Nuclear Reactor Safety	$ (30.6)	$ (47.1)
Nuclear Materials & Waste Safety	124.0	127.7
Net Cost of Operations	$ 93.4	$ 80.6

Analysis of the Statement of Net Cost

The Statement of Net Cost presents the net cost of NRC's two programs as identified in the NRC Annual Performance Plan. The purpose of this statement is to link program performance to the cost of programs. The NRC's net cost of operations for the year ended September 30, 2007, was $93.4 million, which is an increase of $12.8 million over the FY 2006 net cost of $80.6 million. Net costs by program are shown in the accompanying table. Gross costs increased primarily due to an increase in Nuclear Reactor Safety in the areas of new reactor and existing licensing programs. Earned Revenue increased primarily because of the increase in appropriations for NRC activities, of which the NRC is required to collect 90 percent through fee billing.

Total earned revenue for the year ended September 30, 2007, was $693.3 million, which is an increase of $53.3 million from the earned revenue of $640.0 million for the year ended September 30, 2006. Earned revenue is derived from fees for reactor and materials licensing and inspections in accordance with 10 CFR Parts 170 and 171.

Analysis of Statement of Changes in Net Position

The Statement of Changes in Net Position reports the change in net position during the reporting period. Net position is affected by changes in its two components—Cumulative Results of Operations and Unexpended Appropriations. The increase in Net Position of $68.6 million from FY 2006 to FY 2007 is due primarily from an increase in the net change in Unexpended Appropriations of $60.3 million. This increase is primarily due to the increase in the appropriation for FY 2007 for the expected added volume of new reactor licensing activities.

Analysis of the Statement of Budgetary Resources

The Statement of Budgetary Resources reports the source and status of budgetary resources at the end of the period. It presents the relationship between budget authority and budget outlays, and the reconciliation of obligations to total outlays. For FY 2007, NRC had Total Budgetary Resources available of $910.9 million, the majority of which was derived from new budget authority. This represents a 13 percent increase over FY 2006 budgetary resources available of $809.0 million. The increase provides funding for the anticipated growth in new reactor licensing including costs for staffing, pre-application activities, and office space.

For FY 2007, the NRC had Obligations Incurred of $838.8 million, or 92 percent of funds available, compared to FY 2006 Obligations Incurred of $734.8 million, at 91 percent of funds available. This increase was due primarily to the increase of appropriations received for new reactor licensing activities. Gross outlays for FY 2007 were $764.4 million, which represents a $77.8 million increase from FY 2006 total outlays of $686.6 million primarily due to the increase in spending in the area of Nuclear Reactor Safety for new reactor and existing reactor licensing programs.

U.S. NUCLEAR REGULATORY COMMISSION
FEDERAL MANAGERS' FINANCIAL INTEGRITY ACT
STATEMENT FOR FY 2007

The U.S. Nuclear Regulatory Commission's (NRC) management is responsible for establishing and maintaining effective internal controls and financial management systems that meet the objectives of the Federal Managers' Financial Integrity Act (FMFIA). The NRC is able to provide a qualified statement of assurance that the internal controls and financial management systems meet the objectives of FMFIA, with the exception of one material weakness noted herein.

The NRC conducted its assessment of the effectiveness of internal control over the effectiveness and efficiency of operations and compliance with applicable laws and regulations in accordance with OMB Circular A-123, Management's Responsibility for Internal Control. Based on the results of this evaluation, the NRC identified one material weakness in its internal control over the effectiveness and efficiency of operations and compliance with applicable laws and regulations as of September 30, 2007. Other than this exception, the internal controls were operating effectively, and no other material weaknesses were found in the design or operation of the internal controls.

In addition, the NRC conducted its assessment of the effectiveness of internal control over financial reporting, which includes safeguarding of assets and compliance with applicable laws and regulations, in accordance with the requirements of Appendix A of OMB Circular A-123. Based on the results of the evaluation, the NRC can provide reasonable assurance that its internal control over financial reporting as of June 30, 2007, was operating effectively, and no material weaknesses were found in the design or operation of the internal control over financial reporting.

Dale E. Klein
Chairman
U.S. Nuclear Regulatory Commission
November 15, 2007

SYSTEMS, CONTROLS, AND LEGAL COMPLIANCE

Management Assurances

This section provides information on the NRC's compliance with the Federal Managers' Financial Integrity Act, OMB Circular A-123, Management's Responsibility for Internal Control, and the Federal Financial Management Improvement Act. A summary of Management Assurances is included in Appendix D.

Federal Managers' Financial Integrity Act

The Federal Managers' Financial Integrity Act (Integrity Act) mandates that agencies establish controls that reasonably ensure that (1) obligations and costs comply with applicable law; (2) assets are safeguarded against waste, loss, unauthorized use, or misappropriation; and (3) revenues and expenditures are properly recorded and accounted for. This Act encompasses program, operational, and administrative areas, as well as accounting and financial management. It also requires the Chairman to provide an assurance statement on the adequacy of internal controls and conformance of financial systems with governmentwide standards.

Management Control Review Program

Managers throughout the NRC are responsible for implementing effective controls in their areas of responsibilities. Each office director and regional administrator prepares an annual assurance statement, which identifies any control weaknesses that require the attention of the NRC's Executive Committee on Internal Control (ECIC). These statements are based on various sources including management knowledge gained from the daily operation of agency programs and reviews, management reviews, program evaluations, audits of financial statements, reviews of financial systems, annual performance plans, Inspector General and Government Accountability Office reports, and reports and other information provided by the Congressional committees of jurisdiction.

The NRC's ECIC is comprised of senior executives from the offices of the Chief Financial Officer and the Executive Director of Operations, with the General Counsel and the Inspector General participating as advisors. The ECIC met and reviewed the assurance statements provided by the offices and regions. The ECIC then informed the Chairman as to whether the NRC had any internal control deficiencies serious enough to be reported as a material weakness or material noncompliance.

The NRC's ongoing internal control program requires, among other things, that internal control deficiencies be integrated into the offices' and regions' annual operating plans. The operating plan process provides for periodic updates and ensures that key issues receive senior management attention. The internal control information in these plans, combined with the individual assurance statements discussed previously, provide the framework for monitoring and improving the agency's internal controls on an ongoing basis.

FY 2007 Integrity Act Results

The NRC evaluated its internal control systems for the fiscal year ending September 30, 2007. This evaluation provided reasonable assurance that the agency's internal controls achieved their intended objectives in accordance with the Integrity Act. The NRC is able to provide a qualified statement of assurance that the internal controls and financial management systems meet the objectives of the Integrity Act, with the exception of one material weakness.

Material Weakness

The Office of the Inspector General performed an independent evaluation of the NRC's implementation of the Federal Information Security Management Act for FY 2007. The following two findings were identified as significant deficiencies in NRC's information technology (IT) security program:

- Only 2 of 30 operational NRC information systems have a current certification and accreditation, and only 4 out of the 11 systems used or operated by a contractor or other organization on behalf of the agency have a current certification and accreditation.

- Annual contingency plan testing is still not being performed for all of the NRC's operational information systems.

As a result of this evaluation, the NRC identified these two findings as one material weakness associated with the agency's overall IT security program under the provisions of the Integrity Act. The NRC will implement the following corrective actions to resolve this material weakness:

- The NRC's FY 2008 budget includes additional resources and the agency has developed a milestone plan to ensure that one-half of the systems will be certified and accredited by September 2008, with the remaining systems being certified and accredited by September 2009.

- All system contingency plan testing will be completed during FY 2008.

OMB Circular A-123, Management's Responsibility for Internal Control, Including Appendix A, Internal Control over Financial Reporting

In FY 2006, the NRC implemented the requirements of the Office of Management and Budget revised Circular A-123, which defined and strengthened management's responsibility for internal control in Federal agencies. The revised Circular included updated internal control standards and a new section, Appendix A, which required Federal agencies to assess the effectiveness of internal control over their financial reporting and prepare a separate statement of assurance as of June 30.

In FY 2007, the NRC continued its assessment of internal control over financial reporting. The scope of financial reports, materiality values, risk assessments, key processes and key controls were re-evaluated. A three-year rotational testing plan was developed, and three of the original nine key processes from FY 2006 were determined to be significant enough to be included in the testing each year of the 3-year cycle. The remaining six key processes will be tested once in the 3-year cycle beginning this year, two each year. Based on the results of this evaluation, the NRC can provide reasonable assurance that its internal control over financial reporting was operating effectively as of June 30, 2007, and that no material weaknesses were found in the design or operation of the internal controls over financial reporting.

Federal Financial Management Improvement Act

The Federal Financial Management Improvement Act (Improvement Act) requires each agency to implement and maintain systems that comply substantially with (1) Federal financial management system requirements, (2) applicable Federal accounting standards, and (3) the standard general ledger at the transaction level. The Improvement Act requires the Chairman to determine whether the agency's financial management systems comply with the Improvement Act and to develop remediation plans for systems that do not comply.

FY 2007 Improvement Act Results

As of September 30, 2007, the NRC evaluated its financial systems to determine if they complied with applicable Federal requirements and accounting standards required by the Improvement Act. The following eight systems were evaluated—the Federal Financial System, Federal Personnel and Payroll system, Human Resources Management System, Cost Accounting System, Advice of Allotments/Financial Plan System, Capitalized Property System, Fee Billing System, and Controller Resource Database System.

As of September 30, 2007, the agency's financial management systems are in substantial compliance with the Improvement Act, except for two systems which are in substantial noncompliance because of FISMA significant deficiencies related to lack of current certification and accreditation. In addition, a general support system, which all financial management systems either reside on or rely on, does not have a current certification and accreditation and did not have the annual contingency plan tested. Although there may be a potential risk with security controls, there are a number of existing mitigating controls that provide NRC management reasonable assurance that the financial data resulting from financial management systems is accurate. In making this determination, the NRC considered all the information available to them, including the report from the NRC Executive Committee on Internal Control on the effectiveness of internal controls, Office of the Inspector General Audit reports, and the results of the agency's financial management systems reviews. The agency also relied on the Department of the Interior National Business Center's (DOI-NBC) annual reasonable assurance statement in which they concluded that, for FY 2007, the cross-serviced financial systems are in substantial compliance with Federal financial management systems requirements.

The Fee Billing System was identified by the independent auditors as an Improvement Act noncompliance in the FY 2004 through FY 2006 Financial Statement Audit. The agency has taken a number of additional remediation actions during FY 2007 to improve quality assurance over license fee billing processes. Mitigating controls are currently in place to resolve the finding of substantial non-compliance related to quality assurance procedures and to reduce the risk that errors will go undetected.

Prompt Payment

The Prompt Payment Act requires Federal agencies to make timely payments to vendors for supplies and services, to pay interest penalties when payments are made after the due date, and to take cash discounts when they are economically justified. In FY 2007, the NRC paid 8,966 invoices that were subject to the Prompt Payment Act. The NRC percentage of on-time payments subject to the Prompt Payment Act for FY 2007 is 95 percent (see Figure 7). The amount of interest penalties incurred during FY 2007 was $11,160.

PROMPT PAYMENT

(Percentage)   **Figure 7**

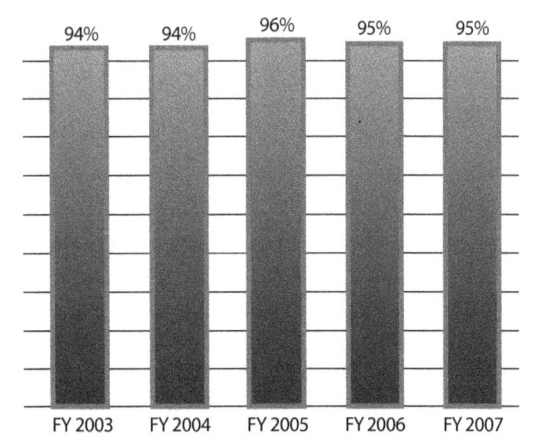

Improper Payments

Improper payments continue to be at low risk for the agency. The NRC continues to evaluate its internal controls to guard against improper payments and monitors and reports on improper payments within its programs. At the present time, NRC's payments consist of commercial vendor, interagency, and

travel reimbursements. The DOI-NBC's Federal Personnel/Payroll System, as the system of record for payroll disbursements, is responsible for monitoring and reporting on any improper payroll-related payments. The NRC continues to perform annual risk assessments for each of these areas. Based on the FY 2007 risk assessments, the number of and amount of improper payments fall below external reporting requirement established by OMB guidance on what is considered to be a significant risk. NRC awards less than $500 million in annual contracts, and, therefore, is not subject to annual reporting under the Recovery Auditing Act.

Debt Collection

The Debt Collection Improvement Act enhances the ability of the Federal Government to service and collect debts. The agency's goal is to maintain the delinquent debt owed to the NRC, at year end, to less than 1 percent of its annual billings. The NRC continues to meet this goal and at the end of FY 2007 delinquent debt was $1.4 million (see Figure 8). The NRC continues to pursue the collection of delinquent debt and refers all eligible delinquent debt over 180 days to the U.S. Treasury for collection.

DELINQUENT DEBT

(In Millions)   **Figure 8**

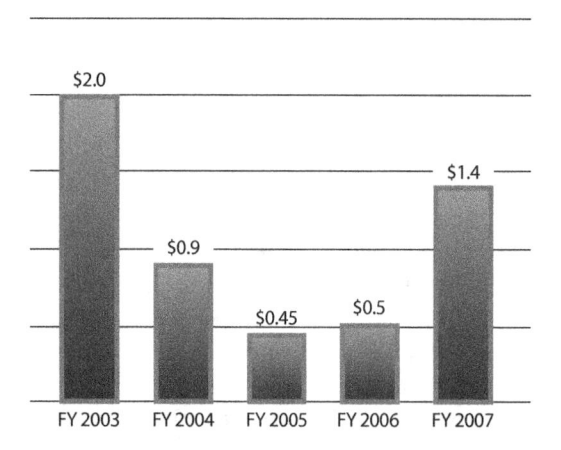

Biennial Review of User Fees

The Chief Financial Officers Act requires agencies to conduct a biennial review of fees, royalties, rents, and other charges imposed by agencies, and make revisions to cover program and administrative costs incurred. Each year, the NRC revises the hourly rates for license and inspection fees and adjusts the annual fees to meet the fee collection requirements of the Omnibus Budget Reconciliation Act of 1990, as amended. The most recent changes to the license, inspection, and annual fees are described in the *Federal Register* (72 FR 31401, June 6, 2007).

The NRC also revised the fees and charges for the Criminal History Program to more appropriately recognize actual costs. Reviews of other types of fees concluded that revisions were not warranted at this time.

Inspector General Act

The agency has established and continues to maintain an excellent record in resolving and implementing open audit recommendations presented in reports from the Office of the Inspector General. Section 5(b) of the Inspector General Act requires agencies to report on final actions taken on audit recommendations. Appendix C includes this information, as well as data concerning disallowed costs determined through contract audits conducted by the Defense Contract Audit Agency.

The Monticello Nuclear Power Plant is located in Monticello, MN, 30 miles northwest of Minneapolis, MN

CHAPTER 2

Program Performance

Fort Calhoun Nuclear Power Plant located between Fort Calhoun, NE and Blair, NE

McGuire Nuclear Station located northwest of Charlotte, NC, on Lake Norman

MEASURING AND REPORTING PERFORMANCE

This chapter presents information on the NRC's performance in achieving its mission and goals during FY 2007. The agency's mission is to ensure adequate protection of the safety and security of the American public in the use of byproduct, source, and special nuclear materials.

The NRC is developing a new Strategic Plan for FY 2008-FY 2013 that determines the agency's long-term strategic direction. The Commission has approved the framework for the draft Strategic Plan. The Performance and Accountability Report reflects the new goal structure proposed in the agency's draft Strategic Plan and reports performance in support of the Safety and Security strategic goals, as well as Openness, Effectiveness and Management which are referred to as operational goals in this report.

This chapter describes the NRC's achievements in accomplishing the two strategic goals. The Safety goal discussion addresses reactor licensing, new reactor licensing, reactor inspection, fuel facilities, material users, high-level waste repository, decommissioning and low-level waste, and spent fuel storage and transportation. The Security goal discussion addresses emergency preparedness and incident response in the nuclear reactor safety and nuclear materials and waste safety programs. In addition, this chapter also describes the agency's progress in achieving greater effectiveness for the five management initiatives identified in the President's Management Agenda. Lastly, this chapter presents information on data sources, data quality, and the completeness and reliability of performance data. This discussion focuses primarily on the NRC's methods for collecting and analyzing data, ensuring data security, and improving the agency's performance measures and the quality of its data during the current reporting period.

GOALS AND PERFORMANCE MEASURES

Safety Goal: Ensure Protection of Public Health and Safety and the Environment

Strategic Outcomes

The NRC has five strategic outcomes associated with the Safety goal that determine whether the agency has achieved its objective to ensure protection of public health and safety as well as the environment:

- No nuclear reactor accidents.

- No inadvertent criticality events.

- No acute radiation exposures resulting in fatalities.

- No releases of radioactive materials that result in significant radiation exposures.

- No releases of radioactive materials that cause significant adverse environmental impacts.

RESULTS: In FY 2007, the NRC achieved all of its Safety goal strategic outcomes.

Performance Measures

The table that follows lists the performance measures and targets for the FY 2007 Safety goal, as stated in the FY 2007 Performance Budget.

FY 2007 SAFETY GOAL PERFORMANCE MEASURES

Measure	2002	2003	2004	2005	2006	2007
1. Number of new conditions evaluated as red by the Reactor Oversight Process is ≤3.	2	1	1	0	0	0
2. Number of significant accident sequence precursors of a nuclear reactor accident is 0.	1	0	0	0	0	0
3. Number of operating reactors with integrated performance that entered the Manual Chapter 0350 process, or the multiple/repetitive degraded cornerstone column or the unacceptable performance column of the Reactor Oversight Program Action Matrix, with no performance exceeding Abnormal Occurrence Criterion I.D.4 is ≤4.	3	2	1	0	0	1
4. Number of significant adverse trends in industry safety performance with no trend exceeding the Abnormal Occurrence Criterion I.D.4 is ≤1.	0	0	0	0	0	0
5. Number of events with radiation exposures to the public and occupational workers that exceed Abnormal Occurrence Criterion I.A is:						
Reactors: 0	0	0	0	0	0	0
Materials: ≤3	0	0	0	1	0	0
Waste: 0	0	0	0	0	0	0
6. Number of radiological releases to the environment that exceed applicable regulatory limits is:						
Reactors: ≤3	0	0	0	0	0	0
Materials: ≤2	4	0	1	0	0	0
Waste: 0	0	0	0	0	0	0

Analysis of Results

1. **Reactor Oversight Process:** The NRC reactor inspection program monitors nuclear power plant performance in three broad areas—reactor safety, radiation safety, and security. Plant performance is analyzed based on a large number of performance indicators and inspection findings. Each nuclear power plant is then categorized into one of four categories—green, white, yellow, or red. Red findings indicate a finding of high safety significance. Results – There are no red performance indicators or findings.

2. **Reactor significant precursors:** The second measure tracks significant precursor events, a statistical measure of risk that determines the likelihood of an event adversely impacting safety. A significant precursor is an event that has a probability of 1 in 1000 (or greater) of leading to substantial damage to the reactor fuel. Results – No significant precursor events have been identified based on screening reviews.

3. **Reactor performance:** The conditions in this measure indicate whether the NRC identifies significant performance issues in a plant during inspections conducted under the reactor oversight program. If any of the conditions in this measure are met, the NRC will take action to ensure that plant safety is improved. Results – The target of less than or equal to four reactors was not exceeded. One reactor met the conditions in this measure during FY 2007. Palo Verde Unit 3 entered the multiple/repetitive degraded cornerstone column because of safety system equipment problems and the licensee was not effective in addressing the problem. NRC inspections identified the issue and brought it to the attention of licensee management for correction. Palo Verde is scheduled for a significant site review in FY 2008.

4. **Reactor safety trends:** This measure tracks trends for several key indicators of industry safety performance. These indicators provide insights into major areas of reactor performance, including reactor safety, radiation safety, and emergency preparedness. Statistical analysis techniques are applied to each indicator to calculate its long-term trend. These trends represent industry averages rather than individual plant performance. Results – No statistically significant adverse trends have been identified in any of the indicators.

5. **Nuclear material radiation exposures:** This measure tracks the number of radiation exposures to the public and occupational workers that exceed Abnormal Occurrence Criterion I.A, which is defined as those events that produce unintended permanent functional damage to an organ or a physiological system, as determined by a physician. This measure tracks both nuclear reactors and other nuclear material users, such as hospitals and industrial users. Results – No radiation exposures in the reactor and materials area exceeded Abnormal Occurrence Criterion I.A.

6. **Nuclear material releases to the environment:** This measure is an indicator of the effectiveness of the NRC's nuclear material environmental programs. Exceeding applicable regulatory limits is defined as a total effective dose equivalent to individual members of the public that is attributable to a licensed user of nuclear materials but does not exceed 0.1 rem in a year, exclusive of dose contributions from background radiation. Results – No nuclear material releases to the environment that exceeded regulatory limits.

Senior Resident Inspector Dan Kimble showing the Region III Regional Administrator James Caldwell the condensate and feedwater control panel in the LaSalle Main Control Room.

INDUSTRY TRENDS

The NRC measures the effectiveness of its Nuclear Reactor Safety program activities based on the continued safe operation of the Nation's nuclear power plants. In order to demonstrate progress in achieving the agency's Safety strategic goal, the NRC compiles data on overall safety performance using several industry-level performance indicators, a number of which are addressed in the following pages. These indicators (except precursor occurrence rate) show significant improvement in the long-term trends for safety performance of nuclear power plants since 1988, the baseline year for the statistical analyses. Plant operating experience data have yielded a steady stream of improvements in the reliability of plant

systems and components, plant operating procedures, training of power plant operators, and regulatory oversight. For ease of viewing, all the figures in this section display data since 1993.

The industry safety indicators are derived through engineering and scientific analyses by the NRC's Office of Nuclear Reactor Regulation and Office of Nuclear Regulatory Research (RES). The analyses of some events for FY 2007 are still ongoing. The performance indicator results are subject to minor variations as licensees submit revisions to the source data and may differ slightly from data reported in previous years as a result of refinements in data quality. The results of these analyses are reported annually to both the Commission and to Congress.

The Industry's Safety Performance Record

SIGNIFICANT EVENTS

per reactor   **Figure 9**

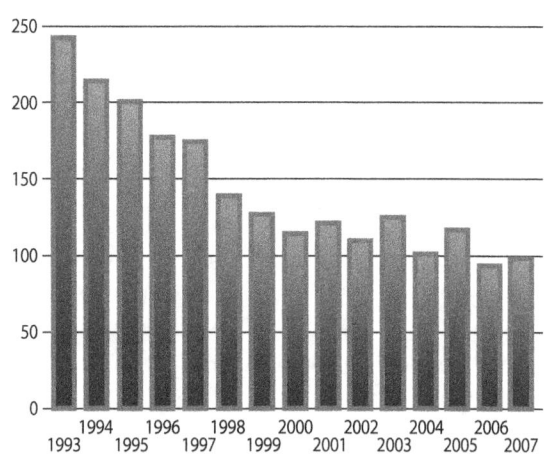

Significant events meet specific criteria such as degradation of important safety equipment. The agency reviews operating events and assesses their safety significance. The number of significant events has declined since 1993.

COLLECTIVE RADIATION EXPOSURE

Exposure (Person-cSv)   **Figure 10**

The total (collective) radiation dose received by workers is an indication of the radiological challenges of maintaining and operating nuclear power plants. The trend shows a reduction in collective dose since 1988 and demonstrates the effectiveness of the controls on radiation exposure implemented to meet these challenges.

SAFETY SYSTEMS ACTUATIONS

per reactor   **Figure 11**

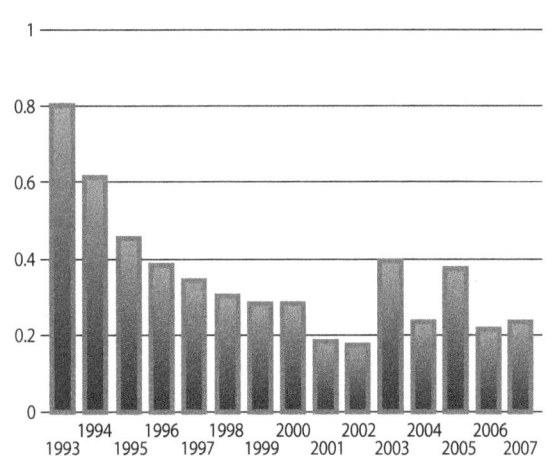

Safety systems mitigate off-normal events such as the widespread power blackout in August 2003, by providing reactor core cooling and water addition. Actuations of safety systems that are monitored include certain emergency core cooling and emergency electrical power systems. Actuations can occur as a result of "false alarms" (such as testing errors) or in response to actual events.

AUTOMATIC SCRAMS

per reactor   **Figure 12**

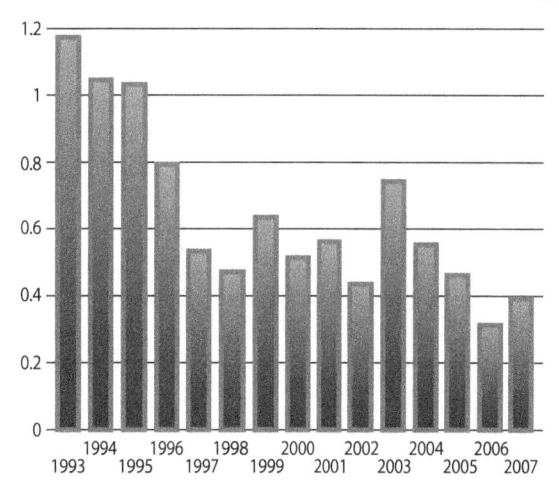

A scram is a basic reactor protection safety function that shuts down the reactor by inserting control rods into the reactor core. Scrams can result from events that range from relatively minor incidents to precursors of accidents. The massive power blackout in August 2003 accounts for most of the increase in FY 2003, but has not affected the statistical trend for number of scrams, which has been declining steadily since 1988.

PRECURSOR OCCURRENCE RATE

per reactor per year   **Figure 13**

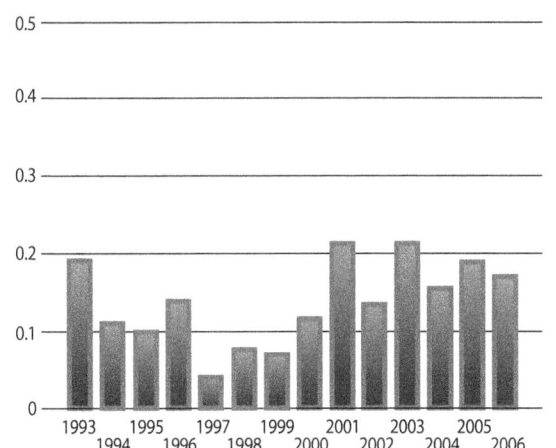

A precursor event is an event that has a probability of greater than 1 in 1 million of leading to substantial damage to the reactor fuel. There is no statistically significant adverse trend in the occurrence rate of precursor events since 1993, the baseline year for the statistical analysis. Due to the complexities associated with evaluating precursor events, the data always lag behind other indicators. Precursor data through FY2007 are not available.

SAFETY SYSTEM FAILURES

per year   **Figure 14**

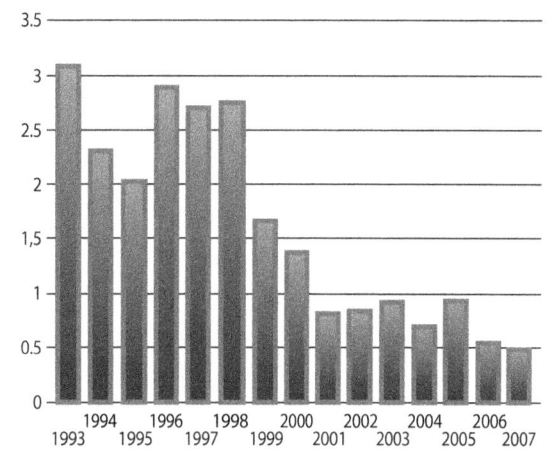

Safety system failures include any events or conditions that could prevent a safety system from fulfilling its safety function. The statistical trend for number of safety system failures across the industry has declined since 1988.

Nuclear Reactor Licensing Activity

The agency's nuclear reactor safety activity ensures that civilian nuclear power reactors and test and research reactors are operated in a manner that adequately protects public health and safety and the environment while safeguarding special nuclear materials used in reactors. Safety at nuclear power plants has improved substantially over the past 20 years, as both the nuclear industry and the NRC have gained extensive experience in the operation and maintenance of nuclear power facilities. This improvement in the safety performance of nuclear power plants results from the combined efforts of the nuclear industry and the NRC.

The NRC completed 1,500 reactor licensing actions during the year (see Figure 15). The agency completed those actions in a timely manner. Approximately 96 percent of the licensing actions in the agency's inventory were completed within 1 year and 100 percent were completed within 2 years (see Figure 16).

LICENSING ACTION TIMELINESS
(Number of Actions)   **Figure 15**

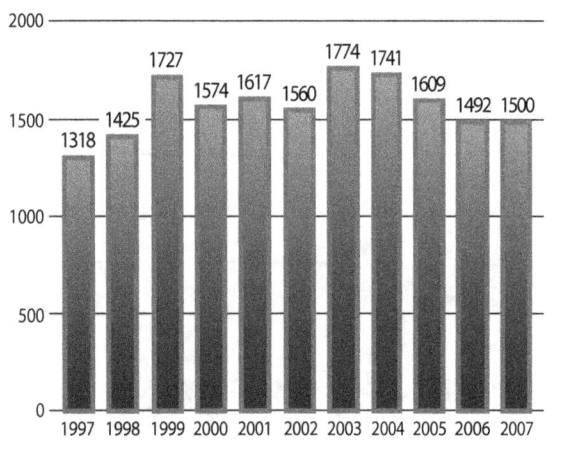

The NRC completed extensive inspection and licensing efforts and authorized the restart of the Tennessee Valley Authority's Browns Ferry Unit 1 nuclear power plant. The NRC staff reviewed

approximately 100 licensing actions and necessary inspections before restart, which took 5 years and approximately 60,000 hours of work.

The NRC also evaluates nuclear reactor power uprate applications, which are means for licensees to increase the power output of their plants. The NRC reviews focus on the potential impacts of the proposed power uprate on overall plant safety and evaluates whether plant operation at the increased power level is safe.

LICENSING ACTION AGE
(In Percent Completed)   **Figure 16**

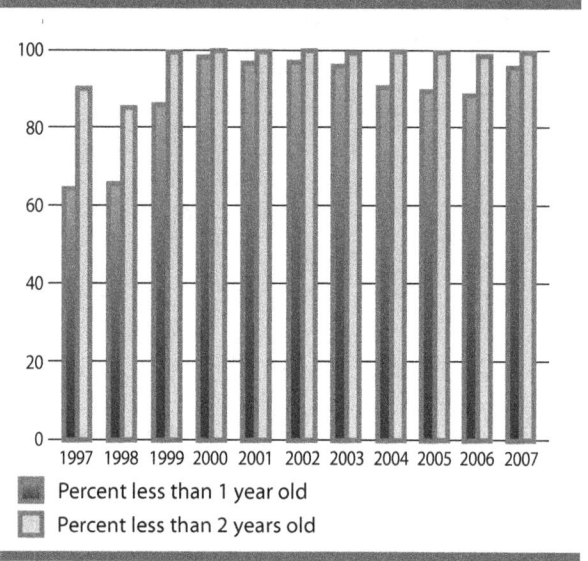

Percent less than 1 year old
Percent less than 2 years old

New Reactor Licensing

Another issue facing the agency is the potential to receive applications for a new generation of nuclear reactors. These licensing activities will ensure that future reactors will meet the NRC's safety requirements and that a stable and predictable regulatory process is in place so that the agency can evaluate future license applications without imposing unnecessary regulatory burden on nuclear power generating companies. The staff is prepared to begin the first application to construct and operate a new reactor which was tendered by NRG Energy for the South Texas site during the last week of September 2007.

The NRC expects to receive a significant number of new reactor combined license applications over the next several years, and continues to develop the infrastructure necessary to support the application reviews. Activities undertaken to prepare for the reviews include issuing a new reactor combined license application regulatory guide (Regulatory Guide 1.206, "Combined License Applications for Nuclear Power Plants [LWR Edition]," issued June 2007), developing strategies for optimizing the review of the applications, developing a construction and vendor inspection program for new construction and vendor activities, and continuing agency activities in the preapplication and design certification review processes. In addition, the NRC has updated more than 250 sections of NUREG-0800, "Standard Review Plan for the Review of Safety Analysis Reports for Nuclear Power Plants," and associated regulatory guides and performed rulemaking activities to revise the licensing process under 10 CFR Part 52. To accomplish these tasks, the NRC reorganized to create the Office of New Reactors.

New Reactor Designs

The NRC has been actively reviewing new nuclear reactor designs to ensure that applications can be evaluated thoroughly and in a timely manner upon receipt. By certifying these designs, the NRC resolves safety issues in a design certification rulemaking. When an applicant submits an application for construction of a new nuclear power plant using one of the certified designs, the license application review can proceed in a manner that promotes safety while minimizing unnecessary regulatory burden and delays for the applicant.

The NRC has issued design certifications for four reactor designs that can be referenced in an application for a nuclear power plant. These designs include the following: General Electric Nuclear Energy's Advanced Boiling Water Reactor design; Westinghouse's System 80+ design; Westinghouse's AP600 design; and Westinghouse's AP1000 design.

The NRC is currently performing the design certification review of the General Electric Economic Simplified Boiling Water Reactor design and is in the process of performing a design certification amendment for the Westinghouse AP1000 design. In addition, the NRC is performing design certification preapplication reviews for the AREVA Evolutionary Power Reactor and Mitsubishi's U.S. Advanced Pressurized-Water Reactor.

Early Site Permits

The NRC has issued early site permits for the Grand Gulf site in Mississippi and for the Clinton site in Illinois. The agency is nearing completion of the early site permit for the North Anna site in Virginia. The staff issued its safety evaluation report for the Vogtle early site permit application on August 30, 2007. Early site permits address site safety issues, environmental protection issues, and plans for coping with emergencies independent of the review of a specific nuclear plant design.

The NRC has revised the regulation governing early site permits, design certifications, and combined licenses (10 CFR Part 52) to improve the effectiveness and efficiency of the licensing process. The NRC published these revised revisions in the *Federal Register* on August 28, 2007.

In addition to working on domestic issues for new reactor construction, the NRC is cooperating with other national nuclear regulatory authorities to address advanced reactor oversight. The NRC is participating in an initiative, the Multinational Design Evaluation Program, through which several regulatory authorities share expertise and resources in reviewing new designs and seek to find ways to harmonize codes, standards, and regulations for the review of future reactor designs and seek to find ways to harmonize codes, standards, and regulations for the review of future reactor designs.

License Renewal

Reactor operating licenses for nuclear reactors are granted for 40 years and can be renewed for an additional 20 years. The review process for renewal applications is designed to assess whether a reactor can continue to be operated safely during the extended period of operation.

To renew a license, the utility must demonstrate that the effects of aging will not adversely affect structures or components important to safety during the renewal period. Such structures and components include the reactor vessel, piping, electrical cabling, containment structure, and steam generators. For some structures or components, additional action may be needed to ensure adequate margins of safety. Additionally, the potential impact on the environment because of the extended period of operation is assessed to verify that the impacts are not so great as to preclude license renewal.

The NRC has received applications to renew the licenses for 62 units at 36 sites and has renewed licenses for 48 units at 26 sites (see Figure 17). The NRC is currently reviewing applications to renew the licenses for 14 units at 10 sites. The agency expects that almost all of the licensees for currently licensed units will ultimately apply to renew their licenses.

LICENSE RENEWAL APPLICATIONS

(Number of Applications)   **Figure 17**

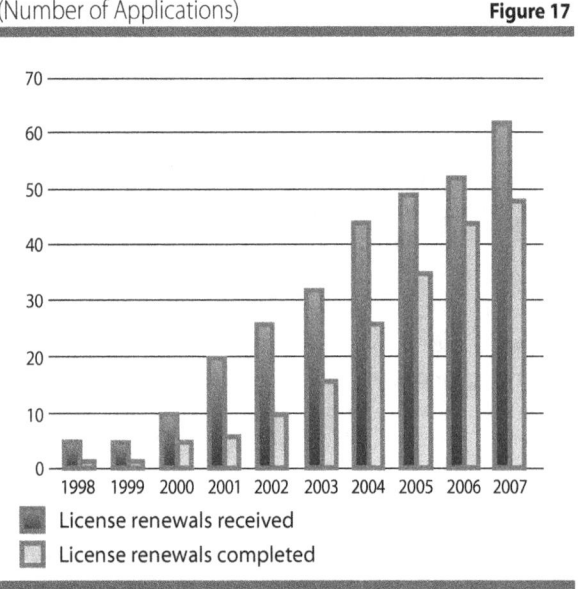

License renewals received
License renewals completed

Nuclear Reactor Inspection

The NRC's Reactor Oversight Process verifies that nuclear plants are being operated safely and in accordance with the NRC's rules and regulations. The NRC has full authority to demand a licensee take immediate action for any conditions that result in excess risk to the public, including requiring a plant to shut down if necessary. The agency evaluates inspection findings and performance indicators to assess the safety performance of each operating nuclear power plant. The NRC performs a rigorous program of inspections at each plant and may perform supplemental inspections and take additional actions to ensure that the plants address significant safety issues. The results of NRC inspection findings for each plant are available to the public at http://www.nrc.gov/NRR/OVERSIGHT/ASSESS/pim_summary.html. The NRC also conducts public meetings with licensees to discuss the results of the NRC's assessments of its safety performance.

In FY 2007, the Nation's nuclear power plants were operated within NRC safety requirements. The performance measures for the Safety goal document that no plants were operating at unacceptable levels. In addition, the safety indicators for nuclear plants as a whole showed no adverse trends. More than 99 percent of plant safety indicators were rated green in FY 2007.

The NRC continued to improve the Reactor Oversight Process in FY 2007. Agency assessments confirm that the Reactor Oversight Process has resulted in a more objective, risk-informed, and predictable regulatory process that focuses NRC and licensee resources on aspects of plant performance that have the greatest impact on safe plant operations.

Reactor Investigations and Enforcement

Compliance with NRC requirements plays an important role in giving the agency confidence that reactor safety is being maintained. NRC policies deter noncompliance and encourage prompt identification and timely comprehensive corrections. Licensees, contractors, and their employees who do not achieve the high standard of compliance expected by the

NRC are subject to enforcement sanctions. Each enforcement action depends on the circumstances of the case. The NRC will not permit licensees to continue to conduct licensed activities if they cannot achieve and maintain adequate levels of safety. In FY 2007, there were 107 escalated enforcement actions with $459,750 in fines assessed in FY 2007.

Fuel Facilities

The NRC licenses and inspects all commercial nuclear fuel facilities that process and fabricate uranium ore into reactor fuel. This fuel is the raw material that powers the Nation's nuclear reactors. Inspection actions include detailed health, safety, safeguards, and environmental licensing reviews as well as inspections of licensee programs, procedures, operations, and facilities to ensure safe and secure operations.

The NRC conducted several significant fuel cycle licensing reviews in FY 2007. The agency completed license renewals for BWX Technologies, Inc., and Westinghouse Electric Co., LLC. To ensure that the fuel facilities are operating safely and securely, the agency reviewed, among other issues, safety analyses for controlling hazardous materials and the engineered and human performance barriers relied on to control hazardous materials. The NRC also conducted comprehensive reviews of fuel cycle licensees. Including an integrated safety analysis which increases the use of risk information to identify hazards, the engineered and human performance barriers relied on to control hazards, and the management measures to ensure that controls are available and reliable. The NRC completed integrated safety analysis reviews for Westinghouse Electric Co., LLC, and AREVA NP, Inc. The NRC also completed a review of the annual integrated safety analysis updates for all fuel facilities.

The NRC issued Orders related to access controls to enhance security at all fuel cycle facilities. The Orders implemented Section 652 of the Energy Policy Act of 2005.

The NRC issued a license to USEC, Inc., to construct and operate the American Centrifuge Plant. This is the second license issued by the NRC for a full-scale uranium enrichment plan. The American Centrifuge Plant will use gas centrifuge technology to enrich uranium. The enriched uranium generated by this facility will provide fuel for nuclear power plants, which will allow the continued safe and secure development of the industry to satisfy the Nation's increasing need for electricity, both now and in the future. Both the American Centrifuge Plant and the National Enrichment Facility, another gas centrifuge facility, are currently under construction.

Nuclear Materials Users

The NRC licenses and inspects the commercial use of nuclear material for industrial, medical, and academic purposes. Commercial uses of nuclear materials include medical diagnosis and therapy, medical and biological research, academic training and research, industrial gauging and nondestructive testing, production of radiopharmaceuticals, and fabrication of commercial products (such as smoke detectors) and other radioactive sealed sources and devices. The NRC and 34 Agreement States regulate more than 21,000 specific materials licensees and 150,000 general materials licensees. The NRC currently regulates and inspects approximately 4,400 specific licensees for the use of nuclear byproduct and other radioactive materials.

Detailed health and safety reviews, as well as inspections of licensee procedures and facilities, provide reasonable assurance of safe operations and the development of safe products. The NRC routinely inspects nuclear materials licensees to ensure that they are using nuclear materials safely, maintaining accountability of those materials, and protecting public health and safety. The agency also analyzes operational experience from NRC and Agreement State licensees. The NRC meets regularly internally to evaluate the safety significance of events reported by licensees and Agreement States.

In FY 2007, the NRC completed reviews of 2,688 materials licensing actions and 1,225 materials program inspections. From 2001 through 2007, the NRC has maintained effectiveness in the timeliness of its reviews of nuclear material license renewals and sealed source and device designs. In FY 2007, the NRC completed 98 percent (109) of the 111 requests for license renewal and sealed source and device design reviews within 180 days, and 98 percent (2,520) of 2,577 new applications and license amendments within 90 days.

The Palisades Senior Resident Inspector (John Ellegood) shows the Region III Regional Administrator (James Caldwell) and the EDO (Luis Reyes) the supplemental Emergency Diesel Generator (EDG). The supplemental EDG provides an additional source of electricity to address concerns over loss of power during a potential station blackout event.

The NRC worked with the Department of Energy to recover unwanted or orphaned radioactive sources. From the inception of this program in 1997, more than 15,500 radioactive sources have been recovered from more than 620 sites within the United States.

The NRC is assisting the Customs and Border Protection agency in fulfilling its congressional mandate to verify the legitimacy of radioactive material shipments coming into the United States through established ports of entry. The NRC regularly provides Customs and Border Protection with

information on the licensing of radioactive materials, including import and export licensing data, and has established processes to provide around-the-clock technical support.

The NRC completed an inventory of high-risk sources, defined as International Atomic Energy Agency (IAEA) Category 1 and Category 2 sources. The NRC also used the inventory in further enhancing the safety, security, and control of radioactive sources, including issuance of increased control orders

In 2005, the NRC issued more than 1,000 increased control orders imposing additional safety and security measures on licensees that possess quantities greater than those specified in IAEA Category 2. The NRC worked with the Agreement States to impose the same requirements on their licensees through legally binding agreements. In addition to continuing in FY 2007 to evaluate the need to enhance security at byproduct material licensees, the NRC is inspecting licensee compliance with these safety and security measures and coordinates with Agreement States to identify and resolve any implementation issues. The NRC also issued security orders to irradiator facilities, manufacturer and distributor facilities, and licensees shipping IAEA Category 1 quantities including orders requiring this group of licensees to implement a program to fingerprint and conduct a criminal history check for access to safeguards information and access to material. The NRC began working with Agreement States to issue orders and legally binding agreements requiring fingerprinting and criminal history checks for access to material to licensees subject to increased controls. The NRC revised its screening process for new license applications to provide increased assurance that the material will be used as intended.

Rulemaking Activities

In FY 2007, the NRC undertook several rulemaking activities to allow the use of radioactive materials while protecting public health and safety and the environment. These activities included publishing several rules that certify the safety of casks for storage of spent nuclear fuel, and implementing a National Source Tracking System for certain sealed sources. The agency also published a rule expanding the definition of byproduct material to include discrete sources of radium-226 and accelerator-produced material.

Investigation and Enforcement

Out of approximately 1,085 inspections, 9 resulted in escalated actions, including the issuance of civil penalties. Violations identified included failure to maintain control over licensed material, comply with requirements of the increased controls order, use two independent methods to secure a portable gauging device to deter/prevent theft, secure licensed material from unauthorized access, and submit accurate information to the NRC. The NRC issued associated civil penalties, including three for $3,250; four for $6,500; one for $9,750; and one for $13,000.

State and Tribal Programs

The NRC, with the assistance of the Agreement States, completed nine Integrated Materials Performance Evaluation Program reviews to determine the adequacy and compatibility of those Agreement States and one review for the materials licensing and inspection program in NRC Region III.

Three States (Nebraska, Massachusetts, and Ohio) signed an addendum that modified their respective Section 274i agreements under the Atomic Energy Act to perform security inspections, for and on behalf of the NRC, of materials licensees authorized to possess and transport items containing radioactive material in quantities of concern.

High-Level Waste Repository

The high-level waste repository activity focused preparing for an application from DOE for permanent storage and disposal of high-level nuclear waste. The NRC conducts its high-level waste program in accordance with the Nuclear Waste Policy Act (as amended), and the Energy Policy Act of 1992.

In FY 2007, the NRC assessed technical and regulatory issues relevant to the proposed repository. The NRC reviewed and evaluated technical and scientific changes to the Department of Energy program; observed and commented on the Department of Energy's quality assurance program; issued enhanced license application review guidance; revised technical models to conform to a new Environmental Protection Agency standard and supplemented, maintained, and operated the Licensing Support Network to allow document access to potential parties to the hearing and the public. The NRC also conducted public outreach activities and meetings to make the regulatory process accessible to interested stakeholders. In addition, the agency provided legal advice, counsel, and representation for staff reviews, Commission actions, and pre-application discovery disputes.

The NRC continued to interact with the Department of Energy on its spent fuel management program, which will use standardized transportation, aging, and disposal canisters. The Department of Energy issued final performance specifications for the disposal container in June 2007, and these specifications will inform the designs for transport package and storage cask systems. These interactions will inform the development of the NRC's approach to reviewing the canister certification application.

To prepare for the eventual high-level waste license application, the NRC enhanced its electronic information exchange capability to enable the electronic receipt of high-level waste documentary material. The agency used the electronic hearing docket in the proceeding for the Prelicense Application Presiding Officer. The NRC obtained security approval to deploy the protective order file to support the proceeding. The NRC tested its preparedness by conducting end-to-end exercises to determine how organizations' processes, procedures, functions, and systems receive, process, and respond to documents and filings. The agency's management group completed the operational readiness review for the release and concluded that the release met the service-level requirements and functionality for the pre-license application phase.

Decommissioning and Low-Level Waste

The NRC licenses and inspects activities at 16 power and early demonstration reactors, 14 research and test reactors, 23 uranium recovery sites, and 32 complex material and fuel cycle facilities that are undergoing decommissioning and the NRC conducts regulatory oversight activities at 16 licensed Title II uranium recovery facilities. Decommissioning removes radioactive contamination from buildings, equipment, ground water, and soil, achieving levels that permit the release of the property, with or without restrictions on its future use by the public. The NRC terminates the licenses for decommissioned facilities after the licensees demonstrate that the residual on-site radioactivity is sufficiently low to protect the health and safety of the public and the environment, and is within regulatory limits. The NRC also conducts a number of regulatory activities to help ensure the safe management and disposal of the low-level radioactive waste generated by radioactive material users, nuclear power plants, and other NRC licensees.

The NRC has overseen decommissioning activities at numerous complex sites and power reactor sites. In FY 2007 the NRC terminated the licenses, or completed regulatory oversight activities, at two power reactors, three research and test reactors, seven complex materials sites, and one uranium recovery site. Completion of decommissioning activities enables sites to return to productive use while ensuring that residual radioactivity does not pose an unacceptable risk to the public.

In FY 2007, the NRC completed monitoring plans for the Savannah River Site (SRS) Saltstone facility and the Idaho National Laboratory (INL) for waste determinations made pursuant to the Ronald W. Reagan National Defense Authorization Act for Fiscal Year 2005 (NDAA). The NRC completed the INL Tank Farm Facility Technical Evaluation Report in October 2006. NRC performed the first on site observation under the NDAA at INL in April 2007. During that April observation at INL, NRC also supported a public meeting with the Snake River Plain

Alliance and other interested members of the public. NRC also worked with DOE, the State, and the EPA to develop an enhanced consultation process for future waste determinations at the SRS. In August 2007, NRC published a Notice of Availability in the *Federal Register* for NUREG-1854, "NRC Staff Guidance for Activities Related to U.S. Department of Energy Waste Determinations, Draft Final Report for Interim Use."

Spent Fuel Storage and Transportation

The NRC ensures that reactor spent fuel is safely stored to support continued reactor operations and safely transported when necessary. The NRC conducts licensing and certification reviews to ensure that storage designs comply with NRC regulations for the storage of nuclear reactor spent fuel and for the domestic and international transport of nuclear reactor spent fuel and other risk-significant radioactive materials.

Shipments of radioactive materials are safely and securely transported each year within the United States. Several Federal agencies share responsibility for regulating the safety and security of those shipments. The NRC closely coordinates its transportation-related activities with those of the Department of Transportation and, as appropriate, the Department of Energy. To help ensure the safety and security of both spent fuel storage and radioactive material transportation, the NRC inspects transport container package designs, spent fuel storage cask designs, and interim storage of spent fuel at both reactor sites and sites away from the reactors.

In FY 2007, the NRC completed 57 transport container design reviews and 10 storage container and installation design reviews (see Figure 18). The NRC review of transportation and interim storage licensing requests ensure that shipments are made in NRC-approved packages that meet rigorous performance requirements and verifies that spent fuel is safely stored, thereby enabling continued reactor operations. The NRC also conducted 14 inspections of independent spent fuel storage installations and radioactive material package certificate holders in order to perform "dry run" loadings with licensee personnel and to ensure that casks are being fabricated according to approved safety requirements.

STORAGE AND TRANSPORTATION DESIGN REVIEWS COMPLETED

Figure 18

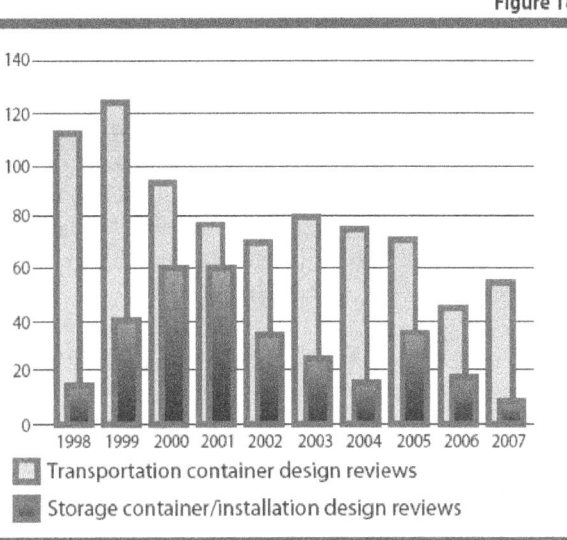

Transportation container design reviews

Storage container/installation design reviews

The NRC issued studies of two tunnel fires (the Baltimore tunnel in Maryland and the Caldecott tunnel in Oakland, California) involving non-nuclear materials to analyze possible regulatory implications of such events for the transportation of spent nuclear fuel. The staff concluded from both evaluations that regulatory requirements for the containment of radioactive material would have been met, and hence the public would be protected from similar events involving radioactive material shipments.

The NRC issued a draft and final supplement for the environmental assessment of the spent fuel storage facility under construction at the Diablo Canyon nuclear plant. The report follows a June 2006 ruling by the U.S. Court of Appeals for the Ninth Circuit that the NRC must consider the possibility of terrorist attacks in reviews of proposed new facilities. The supplemental environmental assessment concludes that the probability of a successful terrorist attack on any such facility is very low. This conclusion is based on the NRC's continual evaluation of the

threat environment and coordination with other Federal, State, and local agencies; protective measures currently in place that reduce the chances of the success of any terrorist attack; the robust design of dry cask storage systems, which provide substantial resistance to penetration; and, the NRC's security assessments of potential consequences of terrorist attacks at these facilities.

Research Activities

Safety Research

The NRC's safety research program evaluates and resolves safety issues for nuclear power plants and other facilities regulated by the NRC, provides the basis for regulatory changes and improvements, develops technical bases and tools to address emerging issues and advanced reactor designs, coordinates NRC activities related to consensus and voluntary standards for agency use, assesses operational events to identify accident precursors, and resolves safety issues. The agency conducts its research program to evaluate existing and potential safety issues; supply independent expertise, information, and technical judgments to support timely and realistic regulatory decisions; reduce uncertainties in risk assessments; and develop technical regulations and standards. When possible, the NRC engages in cooperative research with other Government agencies (e.g. the Department of Energy and the National Aeronautics and Space Administration), the nuclear industry, universities, and international partners.

During the past year, the NRC research program has addressed key areas that support the agency's safety mission, including verification and validation of fire safety models for nuclear power plant applications, development of a licensing strategy for the next-generation nuclear plants, a proactive material degradation assessment of reactor system and pressure boundary components and their susceptibility to known and potential degradation mechanisms, and research to support the licensing of new digital instrumentation and control (I&C) systems.

Fire Safety

The NRC's fire safety research program supports regulatory activities related to fire protection and fire risk analysis. During FY 2007, this research program focused on risk-informed fire protection activities such as supporting the implementation of a new fire protection rule, 10 CFR 50.48(c), which endorses National Fire Protection Association Standard 805, and the fire protection inspection significance determination process. In May 2007, the NRC issued NUREG-1824, "Verification and Validation of Selected Fire Models for Nuclear Power Plant Applications," which documents the verification and validation of five fire modeling tools commonly used in nuclear power plant applications. The NRC completed fire testing and issued NUREG/CR-6931, "Cable Response to Live Fire (CAROLFIRE)," Vols. 1 and 2, on June 1, 2007, which provides research results on cable configurations that were identified as needing further study in and provides the necessary data to develop a cable response model to reduce the uncertainty in predicting electrical cable damage when performing fire modeling analysis.

Licensing of New Nuclear Plants

The Energy Policy Act of 2005 specifies that the Secretary of Energy shall establish the new nuclear plant project. This project consists of research, development, design, construction, licensing, and operation of a prototype nuclear plant, including a very-high-temperature reactor, which can be used to generate electricity, hydrogen, or both. In addition, the Energy Policy Act provides that the NRC shall have licensing and regulatory authority for any reactor authorized under the Act. The Secretary of Energy and the NRC Chairman must jointly develop and submit a licensing strategy for the prototype reactor by August 2008. The NRC has initiated work to develop the licensing strategy discussed in the Energy Policy Act. Toward that end, the NRC and the Department of Energy staff reviewed different licensing strategies and identified the advantages and disadvantages of each with respect to meeting the Congressional mandate of building a prototype by 2020. In addition, the staff convened a group of experts to identify research needed to develop the technical basis for NRC decisions to license a next-generation nuclear plant.

Materials Degradation

The NRC is conducting research on materials degradation to identify susceptible materials and components in light-water reactors. In February 2007, the NRC issued NUREG/CR-6923, "Expert Panel Report on Proactive Materials Degradation Management." Other ongoing activities include (1) evaluating the effectiveness of in-service inspection techniques and programs to detect degradation in components with a high likelihood for degradation, (2) estimating probabilities of failure and associated uncertainties for these components, and (3) performing risk assessments of components that are likely to degrade to evaluate their impact on safety. In May, 2007 the NRC issued a report entitled, "Probabilistic Fracture Mechanics Evaluation of Selected Passive Components." Currently, the NRC is also cooperatively developing and implementing an international research program to address potential future degradation by taking mitigating actions, performing effective and timely inspections, and monitoring and repairing affected components.

Digital Instrumentation and Control

The NRC expects a substantial increase in the use of digital systems for both new reactors and retrofits in operating reactors. As a result, the NRC is updating applicable licensing criteria and regulatory guidance and performing research to support licensing of these new digital instrumentation and control systems. The comprehensive Digital System Research Program Plan defines the instrumentation and control research programs that support the regulatory needs of the agency. The NRC's research will result in the development of licensing review and acceptance criteria for issues such as electrical and communication separation and independence between safety-related and nonsafety-related displays and controls and redundant safety channels (interchannel communications). In addition, the NRC is applying its diversity and defense-in-depth policy as a means to address common-cause failures in digital safety systems. Furthermore, the NRC is actively engaged in ongoing cyber research to ultimately provide regulatory guidance and tools for evaluating digital systems for cyber vulnerabilities, including potential vulnerabilities arising from safety and non-safety system interconnections.

State-of-the-Art Reactor Consequence Analysis

The NRC is developing a best estimate of the off-site consequences from hypothetical severe accidents for operating commercial nuclear power plants to provide the public more realistic information regarding the risk associated with commercial nuclear power plants. The NRC is updating previous consequence studies, performed more than 20 years ago, to base the studies on current information.

For more than 20 years, utilities have been improving their plant designs and operations, inspection methods, operator training, and emergency preparedness. These changes have significantly improved nuclear power plant safety. Over the same period, the NRC, the U.S. nuclear industry, and the international nuclear communities performed extensive severe accident research to understand better the phenomena of severe accidents; the performance of the plants' systems and components under these conditions; the timing, magnitude, and composition of the fission product release; and the effectiveness of the different design and mitigative measures, including emergency preparedness.

International Activities

The NRC's international responsibilities involve participation in activities that support U.S. Government compliance with international treaties and agreements. The NRC is also involved in programs of bilateral nuclear cooperation and assistance and actively supports multinational efforts, such as those sponsored by IAEA and the Organization for Economic Cooperation and Development's Nuclear Energy Agency. One notable accomplishments include the NRC's approval of the Memorandum of Cooperation on Nuclear Safety for the Westinghouse Advanced Pressurized Reactor (AP1000) with the National Nuclear Safety Administration of the People's Republic of China. This memorandum will serve as the basis for cooperation through technical assistance, training, and the sharing of information on the AP1000 Reactor.

The NRC has been a leader in developing and implementing programs focused on leveraging the knowledge and resources within the international regulatory community in the licensing of new reactor designs. The NRC is participating in an initiative, the multinational design evaluation program, through which several regulatory authorities share expertise and resources in reviewing new and future reactor designs.

Since the terrorist attacks on September 11, 2001, the NRC has worked both domestically and internationally to enhance nuclear safety and security through the regulatory oversight of radioactive sources. During FY 2007, the NRC provided assistance for strengthening safety and security oversight of radioactive sources to the regulatory authorities of Armenia, Azerbaijan, Georgia, Iraq, Kazakhstan, Kyrgyzstan, Tajikistan, and Uzbekistan. This assistance focused on developing a national registry of radioactive sources and drafting related laws and regulations.

SECURITY GOAL: Ensure the Secure Use and Management of Radioactive Materials

Strategic Outcome

The NRC has the following strategic outcome associated with the agency's goal to ensure the secure use and management of radioactive materials:

- No instances in which licensed radioactive materials are used domestically in a manner hostile to the security of the United States.

RESULTS: In FY 2007, the NRC achieved its security goal strategic outcome.

Performance Measures

The table below lists the performance measures and targets for the FY 2007 Security goal, as stated in the FY 2007 Performance Budget. The NRC met all of the FY 2007 Security goal performance measure targets.

FY 2007 SECURITY GOAL PERFORMANCE MEASURES

Measure	2002	2003	2004	2005	2006	2007
1. Number of unrecovered losses or thefts of risk-significant radioactive sources is 0.	0	0	0	0	0	0
2. Number of substantiated cases of theft or diversion of licensed, risk-significant radioactive sources or formula quantities of special nuclear material, or attacks that result in radiological sabotage is 0.	0	0	0	0	0	0
3. Number of substantiated losses of formula quantities of special nuclear material or substantiated inventory discrepancies of formula quantities of special nuclear material that are judged to be caused by theft or diversion or by substantial breakdown of the accountability system is 0.	0	0	0	0	0	0
4. Number of substantial breakdowns of physical security or material control (i.e., access control containment or accountability systems) that significantly weaken the protection against theft, diversion, or sabotage is less than 1.	0	0	0	0	0	0
5. Number of significant unauthorized disclosures of classified and/or safeguards information is 0.	0	0	0	0	0	0

Analysis of Results

1. **Unrecovered losses or thefts:** This measure includes any loss or theft of radioactive nuclear sources that the NRC has determined to be risk significant. The measure tracks the NRC's performance in ensuring that those radioactive sources that the agency has determined to be risk significant for the public health and safety are accounted for at all times. The agency used a thorough, detailed, scientific methodology and the public rulemaking process to determine which sources are important. There was no loss or theft of radioactive nuclear material that the NRC determined to be risk significant during FY 2007.

2. **Thefts or diversion:** This measure includes whether NRC-licensed facilities maintain adequate protective capabilities to prevent theft or diversion of nuclear material or sabotage that could result in

harm to the public health and safety. There were no substantiated cases of theft or diversion of licensed, risk-significant radioactive sources or formula quantities of special nuclear material or attacks that resulted in radiological sabotage during FY 2007.

3. **Loss or inventory discrepancy:** This measure includes whether special nuclear material is accounted for at all times and that no losses of this material occur that could lead to the creation of an improvised nuclear device or other type of nuclear device. Furthermore, the measure tracks whether the systems in place at NRC-licensed facilities maintain accurate inventories of special nuclear material that the facilities process, use, or store. There were no substantiated losses of formula quantities of special nuclear material or substantiated inventory discrepancies of formula quantities of special nuclear material that were

caused by theft or diversion or by substantial breakdown of the accountability system during FY 2007.

4. **Substantial breakdowns of physical security:** This measure includes any breakdowns in access control, containment, or accountability systems that significantly weakened the protection against theft, diversion, or sabotage for nuclear materials that the Commission has determined to be risk significant. There were no substantial breakdowns of physical security during FY 2007.

5. **Significant unauthorized disclosures:** This measure includes significant unauthorized disclosures of classified and/or safeguards information that cause damage to national security or public safety. This measure tracks whether information that can harm national security (classified information) or cause damage to the public health and safety (safeguards information) has been stored and used in such a way as to prevent its disclosure to the public, terrorist organizations, other nations, or personnel without a need to know. There were no significant disclosures that caused damage to national security or public safety during FY 2007.

Security Activities

Security Inspections

The NRC maintained vigilant oversight of security in the nuclear industry. During FY 2007, the NRC continued to implement the security cornerstone of the Reactor Oversight Process and completed a comparison of the effectiveness and efficiency of the agency's revised significance determination process and an industry-developed alternative. The comparison effort identified difficulties in using the industry alternative, which resulted in the NRC and industry agreeing to discontinue the assessment of the industry alternative and to use the agency's Significance Determination Process while continuing to address areas needing further clarification. Routine security inspections required by the reactor inspection

program continued, following their commencement in FY 2006. In addition, the NRC completed all inspections to determine whether licensees have adequately accounted for and controlled the spent fuel in their spent fuel pools. The inspections showed that the current programs of all licensees' are adequate to control and account for special nuclear material and that past program deficiencies have been corrected.

Force-on-Force Inspections

The NRC regularly carries out force-on-force inspections at commercial operating nuclear power plants as part of its comprehensive security program. These inspections are used to evaluate and improve the effectiveness of plant security programs to prevent radiological sabotage. The agency's force-on-force inspection program is conducted at least once every three years at each commercial nuclear power plant and fuel facility.

Force-on-force inspections assess a nuclear plant's ability to defend against the design-basis threat, which characterizes the adversary against which plant owners must design appropriate defenses, such as physical protection systems and response strategies. A full force-on-force inspection, spanning two weeks, includes both tabletop drills and simulated combat between a mock commando-type adversary force and the nuclear plant security force. During the attack, the adversary force attempts to reach and damage key safety systems and components that protect the reactor's core (containing radioactive fuel) or the spent nuclear fuel pool, potentially causing a radioactive release to the environment. The nuclear power plant's security force seeks to stop the adversaries from reaching the plant's equipment and causing such a release. In FY 2007, the agency completed 23 force-on-force inspections and submitted its second annual Report to Congress on the results of the NRC security inspection program.

Security Rulemaking

During FY 2007, the NRC undertook security rulemaking activities to promote greater stability of the security requirements placed upon its licensees. The agency proposed revisions to the requirements for fitness-for-duty and access authorization, published a final rule revising the design-basis threats, and published a proposed rule for Nuclear Materials Management and Safeguards System database reporting. The agency also implemented interim fingerprinting requirements in accordance with Section 652 of the Energy Policy Act of 2005.

In addition, the agency made significant progress in the development of security infrastructure for new reactor licensing, including development of the standard review plans for early site permits, design certification, and combined operating licenses; security assessment format and content guides; security requirements during construction; and completion of a memorandum of understanding for consulting with the Department of Homeland Security on new reactor applications. The NRC also completed its security review for the design certification of the General Electric ESBWR and provided technical support for a draft COL regulatory guide; and completed its security review of the ESP for Vogtle.

The NRC continued to improve and formalize its working relationships with external Federal agencies. These activities included the development of a memorandum of agreement between the NRC and the Department of Energy on the harboring of transport vehicles at NRC-licensed sites. The agency recognizes the importance of a coordinated approach to security among the agencies in the Federal Government charged with homeland security responsibilities.

Control of Radioactive Sources

In FY 2007, the NRC maintained its efforts to identify and mitigate the risk of terrorist threats through enhanced security and controls for the use, storage, and transportation of byproduct material and spent nuclear fuel. In collaboration with the Department of Homeland Security, Department of Energy, and other Federal, State, and local agencies, the NRC continued

to assess the potential use of risk-significant sources in radiological dispersal devices and to coordinate efforts to consistently enhance radioactive source protection and security.

The NRC worked with Agreement States to issue new requirements to licensees that enhance the security and control for risk-significant radioactive material. This included development of an inspection program to verify the implementation of these measures. The NRC also completed activities for a final rule to establish the regulatory foundation for the National Source Tracking System, a database for tracking radioactive sources of concern. The rule would require the NRC and Agreement State licensees to report transactions involving the manufacture, transfer, receipt, and disposal of nationally tracked sources (i.e., Category 1 and 2 sources from the IAEA Code of Conduct for the Security of Radioactive Sources). In response to two GAO reports recommending the development of a better tracking system for radioactive sources, the first stage for a National Source Tracking System involved the implementation of a source registry and the development of an interim database. In response to a GAO investigation on the ease of obtaining a new license for radioactive sources, the NRC and Agreement States have implemented a process to screen new license applications or applicants to determine, with reasonable assurance, that the requested materials will be used as intended.

The NRC continued its significant participation in implementing portions of the IAEA Code of Conduct on the Safety and Security of Radioactive Sources, as well as its participation in IAEA committees that are developing guidance documents for the security of radioactive sources during use, storage, and transport. The NRC's involvement in these committees enhances security and public safety and contributes to international and domestic regulatory stability. Under the new export and import regulations that became effective in FY 2006 which impose more stringent controls over the Category 1 and 2 materials defined by the Code of Conduct. The NRC issued 158 licenses for export/import. The NRC is also developing plans to expand the National Source Tracking System to include Category 3 sources.

In FY 2007, the agency also ordered additional security measures at the Louisiana Enrichment Services, National Enrichment Facility; completed an initial security review of the facility; and accredited the facility for the storage of national security information. The agency conducted other information security reviews, including an initial facility security review for a Westinghouse Electric Company facility in Pennsylvania and the Portsmouth Gaseous Diffusion Plant in Ohio.

The agency conducted an operational readiness review of the General Electric-Separation of Isotopes Laser Excitation facility in Wilmington, North Carolina. The NRC issued a classified facility clearance and effected the transfer of classified documentation and components from Australia to General Electric-Separation of Isotopes by Laser Excitation facility in accordance with the provisions of the Agreement for Cooperation between Australia and the United States concerning technology for the separation of isotopes by laser excitation.

Scott Atwater, Region IV DNMS inspector, checking dry fuel storage casks at Arkansas Nuclear One in Russellville, AK.

Spent Fuel

In FY 2007, the agency completed three security plan reviews for proposed independent spent fuel storage installations and issued four security orders to new independent spent fuel storage installations licensees. The NRC also reviewed and approved five spent fuel transportation routes. The agency has been involved with the evaluation of the security measures being developed by Private Fuel Storage, a proposed independent spent fuel storage installation located in Skull Valley, Utah, on the Goshute Indian Reservation. The agency's safety evaluation report concluded that the proposed security measures identified for the Private Fuel Storage facility will provide adequate protection of public health and safety.

Emergency Preparedness and Incident Response

The NRC emergency preparedness and incident response activities ensure that the agency is capable of responding effectively to events at its licensees' sites and that adequate protective measures can and will be taken to mitigate plant damage and to minimize radiation doses to members of the public.

In FY 2007, the NRC worked with States to address replenishment of potassium iodide supplies as a supplement to public protective action plans within the 10-mile emergency planning zones around nuclear power plants; worked with the Department of Health and Human Services to distribute pediatric liquid potassium iodide to States that requested it; and, through its role on the Federal Radiological Preparedness Coordinating Committee, is assisting The White House Office of Science and Technology Policy in its statutory responsibility associated with Public Law 127(f) regarding distribution of potassium iodide 10 to 20 miles outside nuclear power plants.

The agency accelerated upgrades to its incident response center, including improved communications and modernization of the Emergency Response Data System. The NRC also began revising emergency preparedness regulations and guidance to address changes in the threat environment and technological and programmatic advancements. Stakeholders, including the public, are actively involved in the revision process. The proactive approach demonstrated by these activities benefits the public by establishing a more robust, effective response framework that can quickly respond to events; coordinating with other Federal, State, and local agencies; and ensuring the protection of public health and safety.

The agency uses different types of exercises to test and demonstrate its incident response and emergency preparedness capabilities. The exercises provide training; test the agency's plans, procedures, and guidance documents; and test and evaluate the headquarters' incident response facility and critical incident response communication capabilities.

In FY 2007, NRC emergency responders participated in 11 exercises at licensee sites, three of which included the full NRC response team. In addition, the NRC participated in two Governmentwide interagency exercises. The NRC also conducted two other performance-based training activities in the form of tabletop drills.

The results of these exercises and tabletop drills include (1) improved relationships and communications between the NRC Headquarters, NRC Regions, the licensees, and the State emergency management organizations; (2) enhanced interactions with other government organizations (e.g., NORAD); (3) testing and implementation of improved training and team format processes; and, (4) improved effectiveness and efficiency of the NRC's Headquarters Operations Center. Following the conduct of each exercise, the NRC completes a comprehensive review of the exercise and collects lessons learned from participants. The lessons learned are used to correct deficiencies identified in the exercise and enhance the efficiency and/or effectiveness of the facility, guidance documentation, or interaction with exercise partners. Six Priority 2 and 83 Priority 3 Lessons

Learned were developed from the pos-exercise critiques of the headquarters response. The NRC closed 11 Priority 1, 94 Priority 2, and 37 Priority 3 Lessons Learned in FY 2007.

Operational Goals and Associated Performance Measures

Below is a description of the agency's Openness, Effectiveness, and Management operational goal performance measures in FY 2007, as well as agency actions taken to correct those measures that were not achieved.

Openness Goal measures not met and corrective actions taken.

2b. The NRC anticipates that meeting and/or exceeding the Federal Agency Mean Score will remain a challenge in FY 2008. However, the Web Content Management System, when implemented, will satisfy the majority of customer concerns. The NRC anticipates that CMS will be implemented late in FY 2008.

2h. To improve the percentage of documents that are released within the required time frame, in the fourth quarter FY 2006, the NRC implemented an agencywide policy regarding a common method of calculating release dates for documents. As a result, there has been an increase in the percentage of documents released to the public within 6 business days. To improve the percentage further and meet the target, the agency will conduct follow-up sessions with offices/regions individually to communicate the agency's policy and the importance of the timely release of information to the public.

2i. While the agency did not meet the target, there has been an increase in the percentage of documents released to the public within 6 business days. We will continue to work with

FY 2007 OPERATIONAL GOALS AND ASSOCIATED PERFORMANCE MEASURES

Measure	2002	2003	2004	2005	2006	2007
Operational Goal: Openness						
1. 90% of surveyed stakeholders that perceive the NRC to be open in its processes.	New measure in FY 2006				N/A	94%
2. 88% of selected openness output measures that achieve performance targets.	New measure in FY 2006				50%	66%
a. 90% of stakeholder formal requests for information receive an NRC response within 60 days of receipt.	New measure in FY 2006				100%	100%
b. The NRC achieves a 72% user satisfaction score for the agency's public Web site greater than or equal to the Federal Agency Mean score based on results of the yearly American Customer Satisfaction Index for Federal Web sites.	New measure in FY 2006				70%	71%
c. Complete 50% of Freedom of Information Act Requests in 20 days (median).	New measure in FY 2006				61%	67%
d. Issue 90% of Director's Decisions under 2.206 within 120 days.	New measure in FY 2006				100%	100%
e. Make 90% of Final Significance Determination Process Determinations within 90 days for all potentially greater than green findings.	New measure in FY 2006				92%	100%
f. 90% of stakeholders believe they were given sufficient opportunity to ask questions or express their views.	New measure in FY 2006				90%	96%
g. At least 90% of Category 1, 2, and 3 meetings on regulatory issues for which public notices are issued at least 10 days in advance of the meeting.	New measure in FY 2006				92%	93%
h. 90% of non-sensitive, unclassified regulatory documents generated by the NRC and sent to the agency's Document Processing Center that are released to the public by the 6th working day after the date of the document.	New measure in FY 2006				63%	75%
i. 90% of non-sensitive, unclassified regulatory documents received by the NRC that are released to the public by the 6th working day after the document is added to the ADAMS main library.	New measure in FY 2006				77%	87%
Operational Goal: Effectiveness						
1. 70% of selected processes deliver efficiency improvements.	New measure in FY 2006				25%	60%
a. 10% reduction in the average enforcement processing time for Handling Discrimination Allegations. Not Achieved	New measure in FY 2006				N/A	0%
b. Eliminate the requirement for license renewal and approve a living license for the two Category III facilities which have been renewed in FY 2006 and FY 2007.	New measure in FY 2006				Not Eliminated	Not Eliminated

FY 2007 OPERATIONAL GOALS AND ASSOCIATED PERFORMANCE MEASURES - Continued

Measure	2002	2003	2004	2005	2006	2007
Operational Goal: Effectiveness—continued						
c. Improve the timeliness of the review process for nuclear power reactor License Termination Plans by at least 30% over 3 years (FY 2006–FY 2008) as compared to the historical average.		New measure in FY 2006			N/A	N/A
d. Reduce resources expended in support of each interagency exercise by 5% while still accomplishing agency goals for each exercise.		New measure in FY 2006			N/A	5%
e. Implement process enhancements to permit improvement of the reactor rulemaking petition timeliness by 5%.		New measure in FY 2006			N/A	5%
f. Achieve an average 5% reduction on license renewal resources for applications completed in FY 2007.		New measure in FY 2006			N/A	5%
2. No more than one instance per program where licensing or regulatory activities unnecessarily impede the safe and beneficial uses of radioactive materials.		New measure in FY 2006			0	0

offices to ensure employees are aware of the importance of ensuring documents are released within 6 business days. We will continue to provide timely reports to offices on document release statistics.

Effectiveness Goal measures not met and corrective actions taken.

1a. Only two discrimination cases were processed during FY 2007 with an average processing time of 236 days. The agency was not able to meet the ten percent reduction in processing time due to the complexity of utilizing alternative dispute resolution (ADR). The direct costs associated with post-investigation ADR are greater than the costs for processing traditional enforcement actions. Efficiencies have been made and continue to be made in the ADR process which should allow the agency to reduce the processing time for future cases.

1b. The Commission has approved a proposal to extend the license term up to 40 years for fuel cycle facilities subject to 10 CFR Part 70, Subpart H. The applicable regulatory infrastructure to support this change is under development. When completed, the next cycle of Category III fuel cycle licensees would receive a 40-year license, based on approval of the licensees' Integrated Safety Analysis. Realistically, a savings would not be realized until FY 2009 or later, and therefore, no efficiency result was realized for FY 2007.

Management Goal measures not met and corrective actions taken.

1b. The agency has experienced a large growth in FTE's within the last year due to the New Reactor Program ramping up to receive applications from licensees to develop and construct new reactors. As a result, additional budget staff was hired to manage the program which resulted in

FY 2007 OPERATIONAL GOALS AND ASSOCIATED PERFORMANCE MEASURES - Continued

Measure	2002	2003	2004	2005	2006	2007
Operational Goal: Management						
1. 70% of selected support processes deliver efficiency improvements. Not Achieved	New measure in FY 2006				50%	0%
a. Percent reduction in time (10% in FY 2006 and 5% in FY 2007) necessary to add or remove employees from drug testing pool. In FY 2007 all employees were included in the drug testing pool, so this measure is not applicable.	New measure in FY 2006				10%	N/A
b. 5% reduction of agency FTE used to develop and submit the FY 2008 and FY 2009 performance budgets.	New measure in FY 2006				0%	12% increase
c. Issue offer letter 80% of the time within 45 work days of the closing date of the announcement.	New measure in FY 2006				67%	31%
2. 70% of selected NRC management programs deliver intended outcomes. Achieved	New measure in FY 2005			60%	80%	100%
a. Infrastructure management program: 80% of activities achieve their targets	New measure in FY 2005			100%	100%	100%
b. Financial Management & Budget and Performance Integration program: 70% of activities achieve their targets	New measure in FY 2005			67%	67%	88%
c. Expanded electronic government program: 75% of activities achieve their targets	New measure in FY 2005			50%	75%	75%
d. Management of Human Capital program: 80% of activities achieve their targets	New measure in FY 2005			80%	100%	80%
e. Internal Communication program: 100% of activities achieve their targets	New measure in FY 2005			100%	100%	N/A

the agency exceeding the target for this measure. However, the Office of the Chief Financial Officer is currently developing a new budget process as directed by the Commission.

1c. The NRC undertook a Lean Six Sigma study during the second quarter of FY 2007 to evaluate the hiring process from the closing date of the announcement to the offer date and develop recommendations to help streamline that process. The agency is currently leading a separate effort to implement the recommendations made by the Lean Six Sigma study workgroup and to develop a plan to assess NRC's progress towards reducing the hiring time frame to meet the 45-day target.

ADDRESSING THE PRESIDENT'S MANAGEMENT AGENDA

Overview

The President's Management Agenda prescribes Governmentwide initiatives to reform the U.S. Government to be more citizen centered, results oriented, and market based and to promote competition rather than stifle innovation. To achieve this goal, the Administration has identified five initiatives to improve Government performance in the areas of (1) strategic management of human capital, (2) budget and performance integration, (3) competitive sourcing, (4) expanded electronic government, and (5) improved financial management. The following describes the response of the NRC to these initiatives and discusses agency accomplishments during FY 2007 in each of the five areas.

Initiative 1: Strategic Management of Human Capital

The NRC's ability to accomplish its mission depends on its highly skilled and experienced workforce. The Commission is proud of the NRC's ranking as the "Best Place to Work" in the Federal Government based on responses to the 2006 Federal Human Capital survey. Going forward, the NRC anticipates growth in new work, especially in reactor licensing reviews, at a time when increasing numbers of experienced staff are eligible to retire and the agency experiences increased competition for staff from the private sector. To address these challenges, the NRC has streamlined recruitment and the review and approval process for relocation and retention.

Through the use of an automated strategic workforce planning tool, the NRC is able to determine what critical skill/knowledge gaps exist and can gear its recruitment and other programs (e.g., grants and fellowships) appropriately. The agency is currently targeting the following fields for aggressive recruitment and staff development—engineering (nuclear, structural, thermal, geotechnical, electrical, environmental, fire protection, and mechanical), security (physical protection, cyber, and network), nuclear physics, health physics, probabilistic risk assessment, digital instrumentation and control, seismology, volcanology, geology, and hydrology.

For the short-run, demand for skilled individuals appears to be already outpacing the available supply. Efforts are underway to increase the talent pool:

1. The NRC provides grants to support courses, studies, training, curricula, and disciplines pertaining to fields that are important to the work of the agency. This important effort is intended to develop the national academic infrastructure necessary to ensure a viable nuclear workforce in the future.

2. The NRC's scholarship and fellowship programs support students pursuing an education in critical skills related to the agency's regulatory mission in exchange for a commitment to work at the NRC.

3. The NRC established and participates in partnership programs with minority institutions of higher education, including historically black colleges and universities, Hispanic-serving institutions, and tribal colleges and universities to enhance their capacity to train students in fields that are critical to the agency's mission.

4. The NRC is also identifying recruitment champions for selected universities to strengthen and develop relationships with diverse student populations.

The NRC's strategic approach to training and development allows the agency to establish priorities and leverage investments to ensure a comprehensive, integrated, competency-based system of staff training. This year, the NRC conducted concurrent Senior Executive Service candidate development programs and offered more frequent leadership potential programs to meet the need for additional supervisory and managerial positions created by the new reactor program and anticipated retirements. The agency also offered executive leadership seminars and leadership training for new supervisors and team leaders.

Initiative 2: Budget and Performance Integration

The NRC continues to make progress in achieving budget and performance integration in accordance with the President's Management Agenda. This progress includes adopting new outcome-based performance measures aligned with the agency's Strategic Plan, accurately monitoring program performance, and integrating performance information with associated costs. To address these initiatives, the NRC has pursued and completed a number of important actions in FY 2007.

Integrating Planning and Budgeting

The NRC's planning, budgeting, and performance management process links the agency's various budget accounts to its safety and security goals and clearly identifies the budgetary resources devoted to them. The agency's budget identifies the alignment of resources to the safety and security goals. The associated output measures closely link to the agency's Safety and Security goals and performance measures.

Budget Formulation Application

The NRC adopted the budget formulation application in FY 2007 to replace an outdated single-user, desktop database. The Web browser, multiuser budget formulation application has increased efficiency by providing agencywide access to budget information, allowing multiple users access to the system, enabling real-time aggregation of entered budget data, and offering more robust reporting capabilities.

Initiative 3: Competitive Sourcing

One of the NRC's corporate management strategies is to acquire goods and services in an efficient manner. To achieve this, the NRC established output measures associated with the implementation of the competitive sourcing initiative under the President's Management Agenda, adopted a performance-based approach to contracting, and posted procurement synopses on the agency's Web site.

The NRC uploaded its Year 2007 Federal Activities Inventory Reform Act inventory in the Office of Management and Budget's Workforce Inventories Tracking System on June 29, 2007. In accordance with NRC's Competitive Sourcing Plan, the NRC has identified potential commercial activities to be studied

to determine which are appropriate for public-private competition. The NRC completed three business case analyses in FY 2007.

The NRC continues to implement performance-based contracting for facility management services, data entry, information technology, and other support services. To give vendors a better understanding of contract requirements, the NRC includes such criteria as measurable performance requirements, quality standards, quality surveillance plans, and provisions for reducing the fee or price when the vendor fails to perform the services, as required. The NRC continues to exceed its target for expending eligible service contracting dollars through performance-based contracting. In addition, the NRC continues to post all required synopses and solicitations for acquisitions valued at more than $25,000 on its external Web site.

Initiative 4: Expanded Electronic Government

The NRC has aligned its information technology investments with the Federal Government's Electronic Government program (e-gov). The e-gov program, a component of the President's Management Agenda, consists of 25 Presidential Priority Initiatives and 9 Line of Business initiatives. Of these 34 initiatives, NRC is engaged in 22 (full partner in 8 initiatives, transitioning to full partnership in 7 initiatives, and monitoring 7 additional initiatives that might benefit NRC).

The NRC uses e-gov services for payroll, security clearance, acquisition support, Governmentwide customer service, and recruitment, and is aligned with the e-records, budget formulation and geospatial programs. Geospatial programs deal with information that can be described in a geographic fashion, e.g., locations of hospitals, schools, nuclear power plants, or information related to road, river or rail systems. NRC uses geospatial information for site location studies and for incident response. NRC is in the process of currently implementing e-travel, e-training, e-authentication, FISMA reporting and training services, and e-rulemaking. NRC is also converting its paper based employee records to OPM's electronic personnel folder. To institutionalize e-gov, NRC has established procedures to avoid information technology investments that would duplicate other Federal electronic government programs. The NRC receives financial and human resource services from the Department of the Interior, a selected shared service provider, and is in process of replacing its core financial systems.

The NRC emphasizes enterprise architecture in its systems development life cycle methodology and has a Project Management Methodology in place. The Project Management Methodology provides full life cycle guidance for the agency, providing guidance for enterprise architecture, CPIC, infrastructure development and life cycle management processes. An Information Technology Senior Advisory Council, comprising senior business managers, plays an integral role in ensuring technology investments align to the agency's mission and goals and in establishing priorities.

Federal Information Security Management Act

In March 2007, the House Committee on Government Reform's Subcommittee on Technology, Information Policy, Intergovernmental Relations, and the Census graded NRC's compliance with FISMA as an "F". NRC has increased efforts to complete the review, testing, and evaluation of major system security plans and authorities to operate. Eight systems were accredited in FY 2007 with the eighth system being the Reactor Program System, which received its authority to operate on September 28, 2007.

The NRC has an effective information technology security awareness training program. All employees are required to complete an online information technology security awareness course, and NRC information systems security officers and other employees and support contractors with significant security responsibilities are required to complete

a more advanced online technical security awareness course. In addition, the NRC maintains an information technology security Web site and provides information to agency employees for the timely awareness of information technology security issues.

Outwardly Facing Systems

The NRC has identified systems that meet the e-gov system criteria for outward facing e-gov systems. These systems are the Code Development System (CDS), Web-Based Licensing System, Electronic Information Exchange (EIE), and the National Source Tracking System. Of note, the EIE program provides for the transmission of digitally signed electronic documents to the NRC over the Internet. Information received in this manner can then be electronically disseminated directly into the agency's information systems. The NRC's electronic information exchange program plays a major role in enabling the agency to meet the Government Paperwork Elimination Act requirement to allow the public the option of transacting business electronically with the agency. The EIE is used to meet authentication requirements.

The EIE handled approximately 97,000 electronic transactions in FY 2007. The majority of those transactions involved receiving and routing digital fingerprints through NRC security personnel to the Federal Bureau of Investigation for security clearances. This procedure reduces the time required for processing from 1-2 weeks to 2 days. The electronic information exchange is also used to transmit licensing and adjudicatory documents to the NRC resulting in shorter processing times and reduced cost.

Information Technology/Information Management Meta-System

To meet the challenges of high level waste, new nuclear power reactor licensing, and E-filing (conducting agency adjudicatory actions electronically), the NRC has integrated several major agency applications, including the Agencywide Documents Access and Management System, Electronic Information Exchange, Electronic Hearing Docket, Digital Data Management System, and Licensing Support Network. The collection of computer applications, information technology infrastructure components (formerly known as the High-Level Waste Meta-System), and business processes is now referred to as the Information Technology/Management Meta-System. NRC completed a requirements analysis targeting implementation of application and infrastructure enhancements, improvement of business processes, and leveraging existing and new information technology while providing a more robust, secure, and integrated environment.

The NRC will continue to validate the Information Technology/Management Meta-System's capability to support the business processes of high-level waste, E-filing, electronic adjudicatory processes, and new reactor licensing processes through performing iterative testing and exercises. The NRC conducted an Operational Readiness Review that resulted in the acceptance of Release 2 of the Information Technology/Management Meta-System to support the High-Level Waste activities and adjudicatory proceedings. The agency used the Information Technology/Management Meta-System in the Vogtle ESP proceeding to perform electronic filing, review, and distribution of adjudicatory documents.

In partnership with the nuclear industry, NRC has successfully streamlined the process for electronic receipt, and online review of Combined Operating License Applications (COLA). All of the stakeholders are now aligned concerning how an electronic COLA will be formatted, packaged, and submitted to the NRC. The enhanced IT components and business process improvements are implemented in the production environment and have been used during the submittal of the latest revision of the AP1000 Design Control Document.

Initiative 5: Improved Financial Management

The agency's goals for improved financial management include providing reliable, transparent, useful, and timely information to stakeholders and for management decision making; maintaining adequate controls; and implementing integrated and flexible systems to meet the agency's reporting needs. This will ensure that NRC's financial assets are adequately protected consistent with risk.

Financial Statements/Reporting

The NRC received an unqualified audit opinion on its FY 2007 financial statements. The agency's independent auditors eliminated the material weakness and the substantial non-compliance findings for the NRC's License Fee Billing System. The agency implemented a number of new and improved controls to include a validation tool which analyzes and reconciles the completeness and accuracy of billing for reactors and material inspections. As a result, the agency has decreased the risk of potential billing errors and further enhanced the control environment.

Also in FY 2007, the NRC completed its second year of implementing the OMB Circular A-123, Appendix A, requirements for assessing internal control over financial reporting. The deficiencies noted during testing were classified as control deficiencies. No material weaknesses were identified. The agency included the results of the assessment in the Federal Managers' Financial Integrity Act Statement of Assurance.

New Financial Management System

NRC is implementing a new core financial management system hosted by a shared service provider using Web-enabled commercial off-the-shelf software. The new system will combine the functionality of the core accounting, license fee billing, cost accounting, allotment/allowance financial plan and the capitalized property systems into a single enterprise-wide system. This systems strategy will result in more efficient transaction processing utilizing electronic workflow management, greater access to information through the use of ad-hoc reporting tools, and improved overall system performance. An integrated financial management system will also improve internal controls by eliminating multiple data transfers between stand alone systems and the resultant manual reconciliations currently performed to ensure data integrity.

Time and Labor System

NRC is implementing a major upgrade to the time and labor system to be hosted by a shared service provider. The NRC plans to leverage the services of a shared service provider to share costs and lower system life cycle costs. The new version will provide significant changes to the functionality of this software including a Web-enabled capability. The time and labor system will have an improved capability to collect information for fee billing, and cost accounting and provide a wider range of management reports. The time and labor system supports issuing the NRC payroll by providing employee time information to the NRC's E-Payroll system, the Federal Personnel/ Payroll System, hosted by the National Business Center, Department of the Interior.

COSTING TO GOALS, PART REVIEWS, AND PROGRAM EVALUATIONS

Costing to Goals

The NRC is working to improve its cost management capabilities to better align its costs with desired outcomes. This year's Performance and Accountability Report presents the full cost of achieving the Safety and Security goals for two of the agency's programs, Nuclear Reactor Safety and Nuclear Materials Safety. The cost of achieving the agency's Safety goal was $728.7 million and the cost of achieving the agency's Security goal was $59.5 million (see Figure 19).

NRC SAFETY AND SECURITY COSTS

(In Millions) **Figure 19**

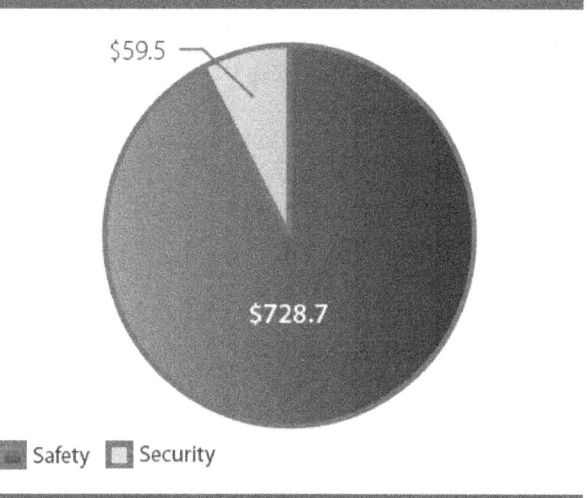

Safety ▇ Security ▇

PROGRAM ASSESSMENT RATING TOOL (PART)

Seven of the agency's major activities have undergone PART reviews, with six of the programs rated as effective, the highest rating available, and one as moderately effective. PART reviews recommended improvement by all programs to develop better linkages between the agency's goals and performance measures. The NRC has responded to this recommendation by defining outcomes and outputs that align with performance measures. The NRC is in the process of linking operating plan performance measures to strategies, outlined in the agency's Strategic Plan, that facilitate the agency meeting its objectives and goals. New measures in the FY 2007 Performance Budget more closely tie the outcomes of the Reactor Inspection and Performance Assessment program to the agency's Safety goal. The following table shows the results of the NRC PART reviews:

PROGRAM	YEAR	PART RATING
Reactor Inspection and Performance Assessment	FY 2003	Effective
Fuel Facilities Licensing and Inspection	FY 2003	Effective
Nuclear Materials Users Licensing and Inspection	FY 2004	Effective
Reactor Licensing	FY 2005	Moderately Effective
Spent Fuel Storage and Transportation Licensing and Inspection	FY 2005	Effective
Decommissioning and Low-Level Waste	FY 2007	Effective
High-Level Waste Repository	FY 2007	Effective

Results of FY 2007 PART Reviews

Decommissioning and Low-Level Waste

Effective in FY 2007. The program earned high scores for Program Purpose and Design and for Program Management. The PART noted that the purpose was clear and the program used regular independent assessments have helped the program to become more results-focused. The program achieves its long-term safety and security goals with respect to the safe management and cleanup of an increasing number of NRC-licensed sites that use radioactive material.

The improvement plan for the program includes developing better linkage of budget requests to the program's success in accomplishing annual and agency long-term goals to make clear how funding affects program accomplishment. Another follow-up action is to improve quantitative measurements of efficiency, including baselines and annual targets to demonstrate year-to-year performance trends better.

High-Level Waste Repository

Effective in FY 2007. The program earned high scores for Program Purpose and Design and for Program Management. The PART noted that the purpose was clear and the program used regular, independent assessments to help the program become more results-focused and to satisfy NRC's Nuclear Waste Policy Act responsibilities and pre-licensing functions. The PART also indicated that the program has made significant progress toward meeting the goal of establishing a regulatory system to ensure the repository achieves long-term safety and security goals.

The improvement plan for the program includes developing better linkage of budget requests to the program's success in accomplishing annual and long-term goals to make clear how funding affects program accomplishment. Another follow-up action is to improve quantitative measurements of efficiency, including baselines and annual targets to demonstrate year-to-year performance trends better.

PROGRAM EVALUATIONS

The NRC conducted a number of self-assessments of its regulatory operations in FY 2007. The license renewal, uranium recovery, and Integrated Materials Performance Evaluation Program activities and the low-level waste program conducted noteworthy evaluations during FY 2007.

License Renewal

The reactor license renewal program evaluation has two objectives: (1) to determine if program elements are effective and efficient and (2) to provide timely, objective information to inform program planning and improvements given the current regulatory environment. The NRC piloted an improved implementation of the license renewal application process for Farley, Arkansas Nuclear One, and D.C. Cook. The primary objective of the program review was to assess the effectiveness of the changes made to the process used by the staff to perform the aging management reviews and aging management program evaluations. The goal of the improved process is to maximize the potential efficiencies available with use of the current license renewal implementation guidance documents by using multi-discipline, on-site review teams. The staff has documented the lessons learned from the pilot application of the improved process and has discussed them in public meetings with stakeholders. Because the improved process was being used on subsequent applications, many of the recommendations before the program review documented them in its report. The agency is currently completing implementation of the remaining recommendations.

NEPA Compliance Assessment/Audit of the Office of Federal and State Materials and Environmental Management Program Environmental Review Program

Battelle Memorial Institute released a comprehensive review of the NRC's Federal and State Materials and Environment Management environmental program, including policies, procedures, and products, and applicable NRC regulations and guidance. The review concluded that the program complies with the National Environmental Policy Act and supports the preparation of documents that fulfill the Act's requirements to concentrate on issues that are truly significant to make the National Environmental Policy Act process useful to decisionmakers and the public. Reviewers identified areas of potential improvement and the staff will incorporate them into the policies, procedures, and practices of the Federal and State Materials and Environment Management environmental review program.

Low-Level Waste Program Strategic Assessment

In FY 2007, an assessment of the NRC's Low-Level Radioactive Waste regulatory program was completed. As part of this assessment, the staff used a variety of means to elicit stakeholder input, including participation in a workshop led by the Advisory Committee on Nuclear Waste and Materials and the issuance of a *Federal Register* notice requesting public comments on the staff's approach. The staff also solicited suggestions from Agreement State regulators and representatives of industry groups and considered recent position papers on Low-Level Radioactive Waste management issued by national scientific and technical organizations.

A strategic objective for the Low-Level Radioactive Waste program was formulated to complement the overall agency goals of Safety and Security. The purpose of the strategic assessment was to identify and prioritize activities to position the Low-Level Radioactive Waste program to meet this objective and to address challenges in areas such as knowledge management. The staff evaluated these activities and assigned priorities of high, medium, or low based on criteria such as their contribution to the agency's strategic goals, the degree of urgency for the activity, and the benefit to be derived. The current Low-Level Radioactive Waste program is fully protective of public health and safety, and the activities evaluated represent opportunities to enhance efforts to risk-inform the Low-Level Radioactive Waste regulatory framework. The staff has begun working on the highest priority activities and will incorporate the results of the assessment into future resource requests.

DATA SOURCES AND QUALITY

The NRC's data collection and analysis methods are driven largely by the regulatory mandate that Congress entrusted to the agency. Specifically, the NRC's mission is to regulate the Nation's civilian use of byproduct, source, and special nuclear materials to ensure adequate protection of public health and safety, protect the environment, and promote the common defense and security. In undertaking this mission, the NRC oversees nuclear power plants, nonpower reactors, nuclear fuel facilities, interim spent fuel storage, radioactive material transportation, disposal of nuclear waste, and the industrial and medical uses of nuclear materials. Section 208 of the Energy Reorganization Act of 1974, as amended, requires the NRC to inform Congress of incidents or events that the Commission determines to be significant from the standpoint of public health and safety. The NRC developed the abnormal occurrence (AO) criteria to comply with the legislative intent of the Energy Reorganization Act to determine which events should be considered "significant." Based on those criteria, the NRC prepares an annual "Report to Congress on Abnormal Occurrences" (NUREG-0090, Vol. 26), which is available on the agency's public Web site at http://www.nrc.gov/reading-rm/doc-collections/nuregs/staff/sr0090.

One important characteristic of this report is that the data presented normally originate from external sources such as Agreement States and NRC licensees. The NRC believes that these data are credible because (1) agency regulations require Agreement States, licensees, and other external sources to report the necessary information; (2) the NRC maintains an aggressive inspection program that, among other activities, includes auditing licensee programs and evaluating Agreement State programs to ensure that they are reporting the information as required by the agency's regulations; and (3) the agency has established procedures for inspecting and evaluating licensees. The NRC employs multiple database systems to support this process, including the Licensee Event Report Search System, the Accident Sequence Precursor Database, the Nuclear Materials Events Database, and the Radiation Exposure Information Report System. In addition, nonsensitive reports submitted by Agreement States and NRC licensees are available to the public through the NRC's Agencywide Documents Access and Management System, which is accessible through the agency's public Web site http://www.nrc.gov.

The NRC has established procedures for the systematic review and evaluation of events reported by the NRC and Agreement State licensees. The NRC's objective is to identify events that are significant from the standpoint of public health and safety based on

criteria that include specific thresholds. The NRC verifies the reliability and technical accuracy of event information reported to the agency. The NRC periodically inspects licensees and reviews Agreement State programs. In addition, NRC headquarters, the Regional offices, and Agreement States hold periodic conference calls to discuss event information. Events identified as meeting the AO criteria are validated and verified by all applicable NRC headquarters program offices, Regional offices, and agency management before being reported to Congress.

Data Security

The NRC ensures data security through its automated information security program, which provides administrative, technical, and physical security measures to protect the agency's information, automated information systems, and information technology infrastructure. Specifically, these measures include the policies, processes, and technical mechanisms used to protect classified information, unclassified safeguards information, and sensitive unclassified information that are processed, stored, or produced on the agency's automated information systems. Data security for information maintained outside the NRC's infrastructure is provided by the hosting contractor or organization.

Performance Data Completeness and Reliability

In order to manage for results, it is essential for the NRC to assess the completeness and reliability of performance data. Comparisons of actual performance with the projected levels are possible only if the data used to measure performance are complete and reliable. Consequently, the Reports Consolidation Act of 2000 requires the Chairman of the NRC to assess the completeness and reliability of the performance data used in this report. In addition, the Office of Management and Budget Circular A-11 specifically describes how Federal agencies should assess the completeness and reliability of their performance data.

Data Completeness

The Office of Management and Budget considers data to be complete if an agency reports actual performance data for every performance goal and indicator in the annual plan. Actual performance data may include preliminary data if those are the only data available when the agency sends its report to the President and Congress. The data presented in this report meet these requirements for data completeness, in that the NRC has reported actual or preliminary data for every strategic and performance goal measure.

Data Reliability

The Office of Management and Budget considers data to be reliable when agency managers and decisionmakers do not demonstrate either a refusal or a marked reluctance to use the data in carrying out their responsibilities. The data presented in this report meets this requirement for data reliability in that the NRC's managers and decisionmakers regularly use the reported data on an ongoing basis in the course of their duties.

Financial Statements and Auditors' Report

A MESSAGE FROM THE CHIEF FINANCIAL OFFICER

I am pleased to present the Nuclear Regulatory Commission's (NRC) Performance and Accountability Report for Fiscal Year 2007. Our independent auditors have once again rendered an unqualified ("clean") opinion for our financial statements, which demonstrates a continued commitment to being good stewards of taxpayers' dollars and ensuring resources are appropriately applied in support of the agency's mission.

As of September 30, 2007, the financial condition of the NRC is sound with respect to having sufficient funds to meet its mission and having adequate control of these funds to ensure our budget authority is not exceeded. As the Chief Financial Officer of the NRC, I take my responsibility for the financial health of the agency very seriously, and I am committed to continuous improvement in our financial management.

Fiscal Year 2007 was a year of great progress for two major initiatives that strengthened financial management. First and foremost, the auditors removed the material weakness finding for the legacy License Fee Billing System. This is primarily due to the extensive effort the agency has made to put in place compensating controls over the past two years that mitigate the risks inherent in the current system. Second, the NRC continued to execute OMB's revised Circular A-123, Appendix A, *Internal Control Over Financial Reporting*. NRC's FY 2007 assessment of Appendix A compliance, while conducted in greater depth with more testing than in FY 2006, again did not identify any material weaknesses. A detailed discussion of the assessment results is included in this report.

Looking to the future, we will continue to be engaged in additional initiatives designed to improve day-to-day operations and achieve long-term strategic planning goals in financial management. The most significant effort entails the replacement of several legacy systems with an integrated, Web-enabled financial management system based on commercial off-the-shelf (COTS) software. The new system will integrate the functionality of a number of current systems including: core accounting, billing, time and labor, cost accounting, and capitalized property systems into a single enterprise-wide system. NRC is currently evaluating and streamlining agency business processes to support a smooth transition to the COTS software. The process changes and replacement systems will improve efficiency and effectiveness while providing agency managers with substantially greater access to timely financial information on which to base their decisions.

To ensure that the NRC's financial assets are adequately protected and reported, the agency's goals for improved financial management include providing reliable, transparent, useful, and timely information to stakeholders and for management decisionmaking; maintaining adequate controls; and implementing integrated and flexible systems to meet the agency's reporting needs. I look forward to the upcoming year to further improve financial management at the NRC, as we make progress in achieving our goals.

William M. McCabe
Chief Financial Officer
November 15, 2007

PRINCIPAL STATEMENTS

BALANCE SHEET
(IN THOUSANDS)

As of September 30,	2007	2006
Assets		
Intragovernmental		
Fund balance with Treasury (Note 2)	$ 356,399	$ 281,715
Accounts receivable (Note 3)	5,228	3,904
Other - Advances and prepayments	3,244	2,247
Total intragovernmental	364,871	287,866
Accounts receivable, net (Note 3)	88,666	71,287
Property and equipment, net (Note 4)	31,832	26,915
Other	39	19
Total Assets	$ 485,408	$ 386,087
Liabilities		
Intragovernmental		
Accounts payable	$ 9,038	$ 8,225
Other (Note 5)	110,797	81,023
Total intragovernmental	119,835	89,248
Accounts payable	18,672	22,940
Federal employee benefits (Note 6)	6,833	7,434
Other (Note 5)	58,877	53,872
Total Liabilities	204,217	173,494
Net Position		
Unexpended appropriations	254,027	193,694
Cumulative results of operations (Note 8)	27,164	18,899
Total Net Position	281,191	212,593
Total Liabilities and Net Position	$ 485,408	$ 386,087

The accompanying notes to the principal statements are an integral part of this statement.

STATEMENT OF NET COST
(IN THOUSANDS)

For the years ended September 30,	2007	2006
Nuclear Reactor Safety		
Gross costs	$ 582,212	$ 515,374
Less: Earned revenue	(612,769)	(562,502)
Total Net Cost of Nuclear Reactor Safety (Note 9)	(30,557)	(47,128)
Nuclear Materials and Waste Safety		
Gross costs	204,495	205,221
Less: Earned revenue	(80,490)	(77,539)
Total Net Cost of Nuclear Materials and Waste Safety (Note 9)	124,005	127,682
Net Cost of Operations	$ 93,448	$ 80,554

The accompanying notes to the principal statements are an integral part of this statement.

STATEMENT OF CHANGES IN NET POSITION
(IN THOUSANDS)

For the years ended September 30,	2007	2006
Cumulative Results of Operations		
Beginning Balance	$ 18,899	$ (13,353)
Budgetary Financing Sources		
Appropriations used	46,646	50,542
Non-exchange revenue	-	590
Transfers-in/out without reimbursement	45,826	45,067
Other Financing Sources		
Imputed financing from costs absorbed by others	27,627	28,022
Other - Revenue from excess collections	(18,386)	(11,415)
Total Financing Sources	101,713	112,806
Net Cost of Operations	(93,448)	(80,554)
Net Change	8,265	32,252
Cumulative Results of Operations	$ 27,164	$ 18,899
Unexpended Appropriations		
Beginning Balance	$ 193,694	$170,836
Adjustment:		
Change in accounting principle (Note 14)	(2,838)	-
Beginning Balance, as adjusted	190,856	170,836
Budgetary Financing Sources		
Appropriations received	109,817	72,532
Appropriations transferred-in/out	-	1,587
Appropriations used	(46,646)	(50,542)
Other adjustments	-	(719)
Total Budgetary Financing Sources	63,171	22,858
Total Unexpended Appropriations	254,027	193,694
Net Position	$ 281,191	$ 212,593

The accompanying notes to the principal statements are an integral part of this statement.

STATEMENT OF BUDGETARY RESOURCES
(IN THOUSANDS)

For the years ended September 30,	2007	2006
Budgetary Resources		
Unobligated balance, brought forward, October 1	$ 74,255	$ 57,349
Recoveries of prior year unpaid obligations		
Actual	5,691	6,642
Budget authority		
Appropriation	824,893	742,686
Spending authority from offsetting collections		
Reimbursements earned - Collected	4,381	6,758
Reimbursements earned - Change in receivables	371	(277)
Change in unfilled customer orders - Advance received	1,433	(2,615)
Change in unfilled customer orders - Without advance	(93)	(358)
Subtotal - Spending authority from offsetting collections	6,092	3,508
Temporarily not available pursuant to public law	-	(461)
Permanently not available	-	(719)
Total Budgetary Resources	$ 910,931	$ 809,005
Status of Budgetary Resources		
Obligations incurred (Note 12)		
Direct	$ 834,126	$ 730,902
Reimbursable	4,645	3,848
Subtotal	838,771	734,750
Unobligated balance		
Apportioned	45,438	48,558
Exempt from apportionment	26,722	25,697
Subtotal	72,160	74,255
Total Status of Budgetary Resources	$ 910,931	$ 809,005
Change in Obligated Balance		
Obligated balance, net		
Unpaid obligations brought forward, October 1	$ 202,446	$ 160,291
Obligations incurred, net	838,771	734,750
Gross outlays	(764,354)	(686,588)
Recoveries of prior year unpaid obligations, actual	(5,691)	(6,642)
Change in uncollected customer payments, from Federal sources	(278)	635
Obligated balance, net, end of period		
Unpaid obligations	274,745	206,019
Uncollected customer payments, from Federal sources	(3,851)	(3,573)
Total unpaid obligated balance, net, end of period	$ 270,894	$ 202,446
Net outlays		
Gross outlays	$ 764,354	$ 686,588
Offsetting collections	(5,814)	(4,143)
Distributed offsetting receipts	(669,245)	(624,042)
Net Outlays	$ 89,295	$ 58,403

The accompanying notes to the principal statements are an integral part of this statement.

NOTES TO THE PRINCIPAL STATEMENTS

Note 1. Summary of Significant Accounting Policies
(All Tables are Presented in Thousands)

A. Reporting Entity

The U.S. Nuclear Regulatory Commission (NRC) is an independent regulatory agency of the Federal Government that was created by the U.S. Congress to regulate the Nation's civilian use of byproduct, source, and special nuclear materials to ensure adequate protection of the public health and safety, to promote the common defense and security, and to protect the environment. Its purposes are defined by the Energy Reorganization Act of 1974, as amended, along with the Atomic Energy Act of 1954, as amended, which provide the foundation for regulating the Nation's civilian use of nuclear materials.

The NRC operates through the execution of its congressionally approved appropriations for salaries and expenses and the Inspector General, including funds derived from the Nuclear Waste Fund. In addition, transfer appropriations are provided by the U.S. Agency for International Development for the development of nuclear safety and regulatory authorities in Russia, Ukraine, Kazakhstan and Armenia for the independent oversight of nuclear reactors in these countries.

B. Basis of Presentation

These principal statements were prepared to report the financial position and results of operations of the NRC as required by the Chief Financial Officers Act of 1990 and the Government Management Reform Act of 1994. These financial statements were prepared from the books and records of the NRC in conformity with accounting principles generally accepted in the United States of America, the requirements of Office of Management and Budget (OMB) Circular A-136, Financial Reporting Requirements, and NRC accounting policies. These statements are, therefore, different from the financial reports, also prepared by the NRC pursuant to OMB directives, which are used to monitor and control NRC's use of budgetary resources.

NRC has not presented a Statement of Custodial Activity because the amounts involved are immaterial and incidental to its operations and mission.

C. Budgets and Budgetary Accounting

Budgetary accounting measures appropriation and consumption of budget spending authority or other budgetary resources and facilitates compliance with legal constraints and controls over the use of Federal funds. Under budgetary reporting principles, budgetary resources are consumed at the time of purchase. Assets and liabilities, which do not consume current budgetary resources, are not reported, and only those liabilities for which valid obligations have been established are considered to consume budgetary resources.

For the past 33 years, Congress has enacted no-year appropriations, which are available for obligation by NRC until expended. For FY 2007, the Revised Continuing Appropriations Resolutions Act 2007 requires the NRC to recover approximately 90 percent of its new budget authority of $824.9 million by assessing fees less amounts derived from the Nuclear Waste Fund of $45.8 million, waste incidental to reprocessing of $2.5 million, and generic homeland security of $33.0 million from P.L. 110-5. The $824.9 million does not include any amounts transferred from the U.S. Agency for International Development.

For FY 2006, NRC recovered approximately 90 percent of its budget authority of $741.5 million less amounts derived from the Nuclear Waste Fund of $45.7 million and waste incidental to reprocessing of $2.5 million.

D. Basis of Accounting

These financial statements reflect both accrual and budgetary accounting transactions. Under the accrual method, revenues are recognized when earned and expenses are recognized when a liability is incurred, without regard to receipt or payment of cash. Budgetary accounting is also used to record the obligation of funds prior to the accrual-based transaction. Interest on borrowings of the U.S. Treasury is not included as a cost to NRC's programs and is not included in the accompanying financial statements.

E. Revenues and Other Financing Sources

The NRC is required to offset its appropriations by the amount of revenues received during the fiscal year from the assessment of fees. The NRC assesses two types of fees to recover its budget authority: (1) fees assessed under 10 CFR Part 170 for licensing, inspection, and other services under the authority of the Independent Offices Appropriation Act of 1952 to recover the NRC's costs of providing individually identifiable services to specific applicants and licensees; and (2) annual fees assessed for nuclear facilities and materials licensees under 10 CFR Part 171. All fees, with the exception of civil penalties, are exchange revenues in accordance with Statement of Federal Financial Accounting Standards No. 7, Accounting for Revenue and Other Financing Sources and Concepts for Reconciling Budgetary and Financial Accounting.

For accounting purposes, appropriations are recognized as financing sources (appropriations used) at the time expenses are accrued. At the end of the fiscal year, appropriations recognized are reduced by the amount of assessed fees collected during the fiscal year to the extent of new budget authority for the year. Collections which exceed the new budget authority are held to offset subsequent years' appropriations. Appropriations expended for property and equipment are recognized as expenses when the asset is consumed in operations (depreciation and amortization).

F. Fund Balance with Treasury

The NRC's cash receipts and disbursements are processed by the U.S. Treasury. The fund balances with the U.S. Treasury are primarily appropriated funds that are available to pay current liabilities and to finance authorized purchase commitments. Funds with Treasury represent NRC's right to draw on the U.S. Treasury for allowable expenditures. All amounts are available to NRC for current use.

G. Accounts Receivable

Accounts receivable consist of amounts owed to the NRC by other Federal agencies and the public. Amounts due from the public are presented net of an allowance for uncollectible accounts. The allowance is based on an analysis of the outstanding balances. Receivables from Federal agencies are expected to be collected; therefore, there is no allowance for uncollectible accounts.

H. Non-Entity Assets

Accounts receivable include nonentity assets of $22 thousand and $5 thousand at September 30, 2007 and 2006, respectively, and consist of miscellaneous penalties and interest due from the public, which, when collected, must be transferred to the U.S. Treasury.

I. Property and Equipment

Property and equipment consist primarily of typical office furnishings, nuclear reactor simulators, and computer hardware and software. The costs of internal use software include the full cost of salaries and benefits from agency personnel involved in software development. The Agency has no real property. The land and buildings in which NRC operates are provided by the General Services Administration (GSA), which charges NRC rent that approximates the commercial rental rates for similar properties.

Property with a cost of $50 thousand or more per unit and a useful life of 2 years or more is capitalized at cost and depreciated using the straight-line method over the useful life. Other property items are expensed when purchased. Normal repairs and maintenance are charged to expense as incurred.

J. Accounts Payable

Accounts payable represent vendor invoices for services received by NRC that will be paid at a later date.

K. Liabilities Not Covered by Budgetary Resources

Liabilities represent the amount of monies or other resources that are likely to be paid by NRC as the result of a transaction or event that has already occurred. No liability can be paid by NRC absent an appropriation. Liabilities for which an appropriation has not been enacted are classified as Liabilities Not Covered by Budgetary Resources. Also, NRC liabilities arising from sources other than contracts can be abrogated by the Government acting in its sovereign capacity.

Intragovernmental

The U.S. Department of Labor (DOL) paid Federal Employees Compensation Act (FECA) benefits on behalf of NRC which had not been billed or paid by NRC as of September 30, 2007, and 2006, respectively.

Federal Employee Benefits

Federal employee benefits represent the actuarial liability for estimated future FECA disability benefits. The future workers' compensation estimate was generated by DOL from an application of actuarial procedures developed to estimate the liability for FECA, which includes the expected liability for death, disability, medical, and miscellaneous costs for approved compensation cases. The liability was calculated using historical benefit payment patterns related to a specific incurred period to predict the ultimate payments related to that period. These projected annual benefit payments were discounted to present value. The interest rate assumptions utilized for discounting benefits were 5.17 percent for both FY 2007 and FY 2006.

Other

Accrued annual leave represents the amount of annual leave earned by NRC employees but not yet taken.

L. Contingencies

Contingent liabilities are those where the existence or amount of the liability cannot be determined with the certainty pending the outcome of future events. The NRC is a party to various administrative proceedings, legal actions, environmental suits, and claims brought by or against it. Based on the advice of legal counsel concerning contingencies, it is the opinion of management that the ultimate resolution of these proceedings, actions, suits, and claims will not materially affect the agency's financial statements. There were no contingent liabilities in FY 2007 or FY 2006.

M. Annual, Sick, and Other Leave

Annual leave is accrued as it is earned and the accrual is reduced as leave is taken. Each year, the balance in the accrued annual leave liability account is adjusted to reflect current pay rates. To the extent that current or prior year funding is not available to cover annual leave earned but not taken, funding will be obtained from future financing sources. Sick leave and other types of nonvested leave are expensed as taken.

N. Retirement Plans

NRC employees belong to either the Federal Employees Retirement System (FERS) or the Civil Service Retirement System (CSRS). For FY 2007 and FY 2006, employees belonging to FERS, the NRC withheld 0.8 percent of base pay earnings, in addition to Federal Insurance Contribution Act (FICA) withholdings, and matched the withholdings with a 11.2 percent contribution in FY 2007 and in FY 2006. The sum is transferred to the Federal Employees Retirement Fund. For employees covered by CSRS, NRC withholds 7 percent of base pay earnings. The NRC matched this withholding with a 7 percent contribution in FY 2007 and FY 2006.

The Thrift Savings Plan (TSP) is a retirement savings and investment plan for employees belonging to either FERS or CSRS. For employees belonging to FERS, NRC automatically contributes one percent of base pay to their account and matches contributions up to an additional four percent. The maximum percentage of base pay that an employee participating in FERS may contribute is unlimited in fiscal years 2007 and 2006. Employees belonging to CSRS may contribute an unlimited percent of their salary in calendar year 2006, but there is no NRC matching of the contribution. The maximum amount that either FERS or CSRS employees may contribute to the plan is $15.5 thousand in calendar year 2007 and $15.0 thousand in the calendar year 2006. The sum of the employees' and NRC's contributions are transferred to the Federal Retirement Thrift Investment Board.

The NRC does not report on its financial statements FERS and CSRS assets, accumulated plan benefits, or unfunded liabilities, if any, applicable to its employees. Reporting such amounts is the responsibility of the U.S. Office of Personnel Management. The portion of the current and estimated future outlays for CSRS not paid by NRC is, in accordance with Statement of Federal Financial Accounting Standards No. 5, Accounting for Liabilities of the Federal Government, included in NRC's financial statements as an imputed financing source.

O. Leases

The total capital lease liability is funded on an annual basis and included in NRC's annual budget. The NRC's capital leases are for personal property consisting of reproduction equipment which is installed at NRC headquarters. For FY 2007, there are seven capital leases with terms of 5 years, consisting of: two capital leases that were added in FY 2007 with an interest rate of 4.58 percent, one capital lease that was added in FY 2006 with an interest rate of 4.25 percent, and four capital leases for FY 2005 with an interest rate of 4.18 percent. The reproduction equipment is depreciated over 5 years using the straight-line method with no salvage value.

Operating leases consist of real property leases with GSA. The leases are for NRC's headquarters and regional offices. The GSA charges NRC lease rates which approximate commercial rates for comparable space.

P. U.S. Department of Energy Charges

Financial transactions between the Department of Energy (DOE) and NRC are fully automated through the U.S. Treasury's Intragovernmental Payment and Collection (IPAC) System. The IPAC System allows DOE to collect amounts due from NRC directly from NRC's account at the U.S. Treasury for goods and/or services rendered. Project manager verification of goods and/or services received is subsequently accomplished through a system-generated voucher approval process. The vouchers are returned to the Office of the Chief Financial Officer documenting that the charges have been accepted.

Q. Pricing Policy

The NRC provides goods and services to the public and other Government entities. In accordance with OMB Circular No. A-25, User Charges, and the Independent Offices Appropriation Act of 1952, NRC assesses fees under 10 CFR Part 170 for licensing and inspection activities to recover the full cost of providing individually identifiable services.

The NRC's policy is to recover the full cost of goods and services provided to other Government entities where (1) the services performed are not part of its statutory mission and (2) NRC has not received appropriations for those services. Fees for reimbursable work are assessed at the 10 CFR Part 170 rate with minor exceptions for programs that are nominal activities of the NRC.

R. Net Position

The NRC's net position consists of unexpended appropriations and cumulative results of operations. Unexpended appropriations represent appropriated spending authority that is unobligated and has not been withdrawn by the U.S. Treasury, and obligations that have not been paid. Cumulative results of operations represent the excess of financing sources over expenses since inception.

S. Use of Management Estimates

The preparation of the accompanying financial statements in accordance with generally accepted accounting principles requires management to make certain estimates and assumptions that directly affect the results of reported assets, liabilities, revenues, and expenses. Actual results could differ from these estimates.

T. Appropriation Transfer

The NRC is a party to allocation transfers with another federal agency (parent) as a receiving (child) entity. Allocation transfers are legal delegations by one agency of its authority to obligate budget authority and outlay funds to another agency. A separate fund account (allocation account) is created in the U.S. Treasury as a subset of the parent fund account for tracking and reporting purposes. All allocation transfers of balances are credited to this account, and subsequent obligations and outlays incurred by the child entity are charged to this allocation account as they execute the delegated activity on behalf of the parent entity. All financial activity related to these allocation transfers (e.g., budget authority, obligations, outlays) is reported in the financial statements of the parent entity from which the underlying legislative authority, appropriations and budget apportionments are derived. The NRC receives allocation transfers, as the child, from U.S. Agency for International Development (USAID). These transfers are for the international development of nuclear safety and regulatory authorities in Russia, Ukraine, Kazakhstan, Georgia and Armenia for the independent oversight of nuclear reactors in these countries.

Note 2. Fund Balance With Treasury

	2007	2006
Fund Balances		
Appropriated funds	$ 301,751	$ 237,956
Allocation transfers	-	3,025
Nuclear Waste Fund	41,300	38,747
Other fund types	13,348	1,987
Total	$ 356,399	$ 281,715
Status of Fund Balance with Treasury		
Unobligated balance		
Available		
Appropriated funds	$ 72,160	$ 74,255
Allocation transfers	-	1,767
Unavailable	-	2,146
Obligated balance not yet disbursed	270,894	203,547
Non-budgetary funds with Treasury	13,345	-
Total	$ 356,399	$ 281,715

Note 3. Accounts Receivable

	2007	2006
Intragovernmental		
Receivables and reimbursements	$ 5,228	$ 3,904
Receivables with the Public		
Materials and facilities fees - billed	$ 2,533	$ 2,094
Materials and facilities fees - unbilled	90,718	72,131
Other	86	109
Total Accounts Receivable	93,337	74,334
Less: Allowance for uncollectible accounts	(4,671)	(3,047)
Accounts Receivable, Net	$ 88,666	$ 71,287

Note 4. Property And Equipment, Net

Fixed Assets Class	Service Years	Acquisition Value	Accumulated Depreciation and Amortization	2007 Net Book Value	2006 Net Book Value
Equipment	5-8	$ 11,957	$ (10,819)	$ 1,138	$ 825
Leased equipment	5-8	1,638	(797)	841	469
IT software	5	47,767	(43,081)	4,686	7,688
IT software under development	-	12,988	-	12,988	5,953
Leasehold improvements	20	25,019	(15,461)	9,558	11,158
Leasehold improvements in progress	-	2,621	-	2,621	822
Total		$ 101,990	$ (70,158)	$ 31,832	$ 26,915

Note 5. Other Liabilities

	2007	2006
Intragovernmental		
Liability to offset net accounts receivable for fees assessed	$ 93,434	$ 75,047
Liability from fees collected which will offset current year's appropriations	13,340	495
Liability to offset miscellaneous accounts receivable	22	5
Liability for advances from other agencies	88	74
Accrued workers' compensation	1,659	1,836
Accrued unemployment compensation	6	15
Employee benefit contributions	2,248	2,060
Liability for clearing account	-	1,491
Total Intragovernmental Other Liabilities	$ 110,797	$ 81,023

The liability to offset the net accounts receivable for fees assessed represents amounts which, when collected, will be transferred to the U.S. Treasury to offset NRC's appropriations in the year collected.

	2007	2006
Accrued annual leave	$ 38,327	$ 35,989
Accrued salaries	15,962	13,815
Contract holdbacks, advances, and other	4,588	4,068
Total Other Liabilities	$ 58,877	$ 53,872

Other liabilities, except accrued annual leave, contract holdbacks, and advances from others are current.

Note 6. Liabilities Not Covered By Budgetary Resources

	2007	2006
Intragovernmental		
FECA paid by DOL	$ 1,659	$ 1,836
Accrued unemployment compensation	6	15
Federal Employee Benefits		
Future FECA	6,833	7,434
Other		
Accrued annual leave	38,327	35,989
Total Liabilities not Covered by Budgetary Resources	$ 46,825	$ 45,274

Balance Sheet amounts represent ending balances of worker's compensation and annual leave.

Note 7. Leases

	2007	2006
Assets Under Capital Leases:		
Copiers and booklet maker	$ 1,638	$ 1,224
Accumulated depreciation	(797)	(755)
Net assets under capital leases	$ 841	$ 469

Future Lease Payments Due:

Fiscal Year	Capital	Operating	2007	2006
2007	$ -	$ -	$ -	$ 23,676
2008	214	23,233	23,447	16,665
2009	206	20,572	20,778	14,373
2010	198	20,638	20,836	14,421
2011	125	20,412	20,537	14,207
2012 and thereafter	107	41,708	41,815	24,419
Total	850	126,563	127,413	107,761
Add: Imputed Interest	86	-	86	32
Total Future Lease Payments	$ 936	$ 126,563	$ 127,499	$ 107,793

Note 8. Cumulative Results of Operations

	2007	2006
Future funding requirements	$ (46,825)	$ (45,274)
Investment in property and equipment, net	31,832	26,915
Contributions from foreign cooperative research agreements	3,184	2,110
Change in Nuclear Waste Fund	38,933	35,107
Other	40	41
Cumulative Results of Operations	$ 27,164	$ 18,899

Future funding requirements represent the amount of future funding needed to pay the accrued unfunded expenses as of September 30, 2007, and 2006. These accruals are not funded from current or prior-year appropriations and assessments, but rather should be funded from future appropriations and assessments. Accordingly, future funding requirements have been recognized for the expenses that will be paid from future appropriations.

Note 9. Statement of Net Cost

The programs as presented on the Statement of Net Cost are based on the annual Performance Budget and are described as follows:

Nuclear Reactor Safety encompasses all NRC efforts to ensure that civilian nuclear power reactor facilities and research and test reactors are licensed and operated in a manner that adequately protects the public health and safety, the environment and protects against radiological sabotage and theft or diversion of special nuclear materials. The Nuclear Reactor Safety program contains three activities – New Reactors, Reactor Licensing and Rulemaking, and Reactor Oversight and Incident Response.

Nuclear Materials and Waste Safety encompasses all NRC efforts to protect the public health and safety and the environment and ensures the secure use and management of radioactive materials. The Nuclear Materials and Waste Safety program contains five activities – Fuel Facilities, Nuclear Materials Users, High-Level Waste Repository, Decommissioning and Low-Level Waste, and Spent Fuel Storage and Transportation.

For "Intragovernmental gross costs," the buyers and sellers are both Federal entities. For "Earned revenues from the public," the buyers of the goods or services are non-Federal entities.

Note 9. Statement of Net Cost (continued)

For the years ended September 30,	2007	2006
Nuclear Reactor Safety		
Intragovernmental gross costs	$ 157,582	$ 147,028
Less: Intragovernmental earned revenue	(36,519)	(32,789)
Intragovernmental net costs	121,063	114,239
Gross costs with the public	424,630	368,346
Less: Earned revenues from the public	(576,250)	(529,713)
Net costs with the public	(151,620)	(161,367)
Total Net Cost of Nuclear Reactor Safety	$ (30,557)	$ (47,128)
Nuclear Materials and Waste Safety		
Intragovernmental gross costs	$ 45,287	$ 48,414
Less: Intragovernmental earned revenue	(7,154)	(6,901)
Intragovernmental net costs	38,133	41,513
Gross costs with the public	159,208	156,807
Less: Earned revenues from the public	(73,336)	(70,638)
Net costs with the public	85,872	86,169
Total Net Cost of Nuclear Materials and Waste Safety	$ 124,005	$ 127,682

Earned revenue for decommissioned reactors was improperly classified under Nuclear Reactor Safety rather than Nuclear Materials and Waste Safety in the prior year. A reclassification has been made to the reported amounts for the first quarter FY 2006 to conform to the current year presentation. Total earned revenue for that period did not change.

	Reported FY 2006	Reclassifications	Reclassified FY 2006
Intragovernmental Earned Revenue			
Reactor Safety	$ (33,121)	$ 332	$ (32,789)
Materials and Waste Safety	(6,569)	(332)	(6,901)
Earned Revenue from Public			
Reactor Safety	(532,661)	2,948	(529,713)
Materials and Waste Safety	(67,690)	(2,948)	(70,638)
Total	$ (640,041)	$ -	$ (640,041)

Note 10. Exchange Revenues

	2007	2006
Fees for licensing, inspection, and other services	$ 687,632	$ 635,457
Revenue from reimbursable work	5,627	4,584
Total Exchange Revenues	$ 693,259	$ 640,041

Note 11. Financing Sources Other Than Exchange Revenue

Appropriated Funds Used

Collections were used to reduce the fiscal year's appropriations recognized:

	2007	2006
Funds consumed	$ 757,892	$ 685,134
Less: Collection from fees assessed	(669,246)	(624,042)
Less: Nuclear Waste Funding Used	(42,000)	(10,550)
Appropriated Funds Used	$ 46,646	$ 50,542

Funds consumed includes $74.3 million and $59.6 million through September 30, 2007, and 2006, respectively, of available funds from prior years.

Non-Exchange Revenue

	2007	2006
Civil penalties	$ 450	$ 461
Miscellaneous receipts	1,681	129
Contra-Revenue	(2,131)	-
Total Non-Exchange Revenue	$ -	$ 590

Imputed Financing

	2007	2006
Civil Service Retirement System	$ 10,593	$ 11,256
Federal Employee Health Benefit	16,956	14,912
Federal Employee Group Life Insurance	71	66
Judgements Awards	7	1,788
Total Imputed Financing	$ 27,627	$ 28,022

Transfers In/Out

	2007	2006
Transfers out to Treasury		
License Fees	$ 69,245	$ 624,042
Non-exchange revenue	-	590
Total Transfers-Out to Treasury	$ 669,245	$ 624,632

Note 12. Total Obligations Incurred

	2007	2006
Direct Obligations		
Category A	$ 788,875	$ 687,201
Exempt from Apportionment	45,251	43,701
Total Direct Obligations	834,126	730,902
Reimbursable Obligations	4,645	3,848
Total Obligations Incurred	$ 838,771	$ 734,750

Obligations exempt from apportionment are the result of funds derived from the Nuclear Waste Fund. Category A Obligations consist of NRC appropriations only. Undelivered orders for the Nuclear Waste Fund are $12.2 million and $9.4 million, Salaries and Expenses $215.0 million and $148.1 million, and the Office of the Inspector General $1.7 million and $1.0 million through September 30, 2007, and 2006, respectively.

Note 13. Nuclear Waste Fund (NWF)

Included in NRC's budget for FY 2007 and 2006 are $45.8 million and $45.7 million, respectively, provided from the NWF. Statement of Federal Financial Accounting Standards No. 27, Identifying and Reporting Earmarked Funds, lists three defining criteria for an earmarked fund. Generally, an earmarked fund is established by law to use specifically identified financing sources only for designated activities, and the statute provides explicit authority to retain current, unused revenues for future use. Also, the law includes a requirement to account for and report on the receipt and use of the financing sources as distinguished from general revenues.

In 1982, Congress passed the Nuclear Waste Policy Act of 1982 (PL 97-425) establishing the Nuclear Waste Fund (NWF) to be administered by the Department of Energy (DOE) (42 U.S.C. 10222). Given the terms of the statute, the NWF clearly meets the definition of an earmarked fund from DOE's perspective, and DOE does indeed report the NWF as an earmarked fund in its Performance and Accountability Report (PAR).

However, to NRC the NWF transfer is a source of financing; its receipt of NWF funds is a use of NWF resources. NRC collects no revenue on behalf of the NWF and has no administrative control over it. Furthermore, the U.S. Treasury has no separate fund symbol for the NWF under NRC's agency location code (ALC). The receipt and expenditure of NWF money is reported to the U.S. Treasury under the NRC's primary Salaries and Expenses Fund (X0200).

Based on these facts, the NWF is not an earmarked fund from NRC's perspective. However, in order to provide additional information to the users of these financial statements, enhanced disclosure of the fund is presented below.

The funding is provided to NRC in FY 2007 and 2006 for the purpose of performing activities associated with DOE's application for a high level waste repository at Yucca Mountain, Nevada. These activities included assistance to DOE with the application, review of the application, the conduct of thorough safety and security evaluations, preparation of the safety evaluation report, initiation of the inspection program, ensuring that the regulation process was made available to stakeholders and the general public, and to provide legal advice and representation for staff reviews and Commission actions.

Note 13. Nuclear Waste Fund (NWF) (continued)

The NWF amounts received, expended, obligated, and unobligated balances as of September 30, 2007, and 2006 are shown in the following:

	2007	2006
Appropriations Received	$ 45,826	$ 45,657
Expended Appropriations	$ 45,640	$ 47,554
Obligations Incurred	$ 45,247	$ 43,701
Unobligated Balances	$ 26,717	$ 25,697

Note 14. Change in Accounting Principle

As discussed in note 1T, the NRC receives allocation transfers from USAID. In prior years, the NRC appropriately reported the proprietary activity related to the allocation transfers on its financial statements. The accompanying FY 2006 financial statements include assets of $3,025 thousand; liabilities of $188 thousand, appropriation transfers of $1,587 thousand and costs of $1,448 thousand related to allocation transfers received from USAID in FY 2006 and prior years.

Effective in FY 2007, OMB Circular A-136, *Financial Reporting Requirements* mandates that a parent entity must report all budgetary and proprietary activity in its financial statements, whether material to a child entity, or not. The effect of this reporting change on prior periods should be reported as a change in accounting principle consistent with SFFAS 21, *Reporting Corrections of Errors and Changes in Accounting Principles*.

The cumulative effect of the change on beginning unexpended appropriations is reported in the accompanying FY 2007 Statement of Changes in Net Position as follows:

Unexpended Appropriations:	
Beginning Balances, October 1, 2006	$ 193,694
Less: USAID Allocation transfers	(2,838)
Restated beginning balance, October 1, 2006	$ 190,856

Note 15. Explanation of Differences Between the Statement of Budgetary Resources and the Budget of the U.S. Government

Statement of Federal Financial Standards (SFFAS) No. 7, *Accounting for Revenue and Other Financing Sources*, requires the NRC to reconcile the budgetary resources reported on the Statement of Budgetary Resources to the prior fiscal year actual budgetary resources presented in the Budget of the U.S. Government and explain any material differences. NRC does not have any material differences between the Statement of Budgetary Resources and the Budget of the U.S. Government. The President's Budget with actual results for NRC has not been published for FY 2007. It is expected to be published on February 5, 2008. The estimates will be available on the NRC web page, and the actuals can be found on the OMB web page.

Note 16. Reconciliation of Net Cost of Operations to Budgetary Resources

OMB Circular No. A-136, Financial Reporting Requirements, dated June 29, 2007, requires agencies to reclassify the Statement of Financing to display in the Notes to the Principal Statements. The disclosure requirements are outlined in SFFAS No. 7.

For the years ended September 30,	2007	2006
Budgetary Resources Obligated		
Obligations incurred (Note 12)	$ 838,771	$ 734,750
Less: Spending authority from offsetting collections and recoveries	(11,783)	(10,150)
Less: Distributed offsetting receipts	(669,245)	(624,042)
Net Obligations	157,743	100,558
Other Resources		
Imputed financing from costs absorbed by others	27,627	28,022
Allocation transfer	-	1,444
Other - Revenue from excess collections	(18,386)	(11,415)
Net Other Resources Used to Finance Activities	9,241	18,051
Total Resources Used to Finance Activities	166,984	118,609
Resources Used to Finance Items not Part of the Net Cost of Operations	(79,278)	(46,809)
Total Resources Used to Finance the Net Cost of Operations	87,706	71,800
Components of the Net Cost of Operations that will not Require or Generate Resources in the Current Period	5,742	8,754
Net Cost Of Operations	$ 93,448	$ 80,554

REQUIRED SUPPLEMENTARY INFORMATION

SCHEDULE OF BUDGETARY RESOURCES
(IN THOUSANDS)

For the year ended September 30, 2007	Salaries & Expenses X0200	Office of Inspector General X0300	Nuclear Facility Fees X5280	Total
Budgetary Resources				
Unobligated balances, brought forward, October 1	$ 73,319	$ 936	$ -	$ 74,255
Recoveries of prior year obligations				
Actual	5,243	448	-	5,691
Budget authority				
Appropriation	154,808	836	669,249	824,893
Spending authority from offsetting collections				
Reimbursements earned - Collected	4,381	-	-	4,381
Reimbursements earned - Change in receivables	371	-	-	371
Change in unfilled customer orders - Advance received	1,433	-	-	1,433
Change in unfilled customer orders - Without advance	(93)	-	-	(93)
Subtotal - Spending authority from offsetting collections	6,092	-	-	6,092
Net Transfers	661,721	7,524	(669,245)	-
Total Budgetary Resources	$ 901,183	$ 9,744	$ 4	$ 910,931
Status of Budgetary Resources				
Obligations incurred (Note 12)				
Direct	$ 824,928	$ 9,198	$ -	$ 834,126
Reimbursable	4,645	-	-	4,645
Subtotal	829,573	9,198	-	838,771
Unobligated balance				
Apportioned	44,892	546	-	43,738
Exempt from apportionment	26,718	-	4	26,722
Subtotal	71,610	546	4	72,160
Total Status of Budgetary Resources	$ 901,183	$ 9,744	$ 4	$ 910,931
Change in Obligated Balance				
Obligated balance, net				
Unpaid obligations, brought forward, October 1	$ 200,962	$ 1,484	$ -	$ 202,446
Obligations incurred, net	829,573	9,198	-	838,771
Gross outlays	(755,376)	(8,978)	-	(764,354)
Recoveries of prior year obligations, actual	(5,243)	(448)	-	(5,691)
Change in uncollected customer payments, from Federal Sources	(278)	-	-	(278)
Obligated balance, net, end of period				
Unpaid obligations	273,489	1,256	-	274,745
Uncollected customer payments, from Federal sources	(3,851)	-	-	(3,851)
Total unpaid obligated balance, net , end of period	$ 269,638	$ 1,256	$ -	$ 270,894
Net outlays				
Gross outlays	$ 755,376	$ 8,978	$ -	$ 764,354
Offsetting collections	(5,814)	-	-	(5,814)
Distributed offsetting receipts	-	-	(669,245)	(669,245)
Net Outlays	$ 749,562	$ 8,978	$(669,245)	$ 89,295

AUDITORS' REPORT

UNITED STATES
NUCLEAR REGULATORY COMMISSION
WASHINGTON, D.C. 20555-0001

OFFICE OF THE
INSPECTOR GENERAL

MEMORANDUM TO: Chairman Klein

FROM: Hubert T. Bell
Inspector General

SUBJECT: RESULTS OF THE AUDIT OF THE UNITED STATES NUCLEAR REGULATORY
COMMISSION'S FINANCIAL STATEMENTS FOR FISCAL YEARS 2007 AND 2006
(OIG-08-A-01)

The Chief Financial Officers Act of 1990, as amended, (CFO Act) requires the Inspector General (IG) or an
independent external auditor, as determined by the IG, to annually audit the United States Nuclear Regulatory
Commission's (NRC) financial statements in accordance with applicable standards. In compliance with this
requirement, this memorandum transmits the following R. Navarro & Associates, Inc. Auditors' Reports:

- Independent Auditors' Report on the Financial Statements,

- Independent Auditors' Report on the Effectiveness of Internal Control over Financial Reporting, and

- Report on Compliance with Laws and Regulations.

Objective of a Financial Statement Audit

The objective of a financial statement audit is to determine whether the financial statements are free of material
misstatement. An audit includes examining, on a test basis, evidence supporting the amounts and disclosures
in the financial statements. An audit also includes assessing the accounting principles used and significant
estimates made by management as well as evaluating the overall financial statement presentation.

R. Navarro & Associates' examination was made in accordance with generally accepted auditing standards,
Government Auditing Standards issued by the Comptroller General of the United States, and Office of
Management and Budget (OMB) Bulletin No. 07-04, Audit Requirements for Federal Financial Statements.
The audit included obtaining an understanding of the internal controls over financial reporting and testing and
evaluating the design and operating effectiveness of the internal controls. Because of inherent limitations in
any internal control, there is a risk that errors or fraud may occur and not be detected. Also, projections of an
evaluation of internal control over financial reporting to future periods are subject to the risk that the internal
control may become inadequate because of changes in conditions, or that the degree of compliance with policies
or procedures may deteriorate.

Results of Audit

The results are as follows:

Financial Statements

- FYs 2007 and 2006 - Unqualified opinion

FY 2007 Internal Controls

- Qualified opinion
- Significant Deficiencies:
  - Information Systems Security Controls (Continuing Material Weakness)
  - Fee Billing System (Significant Deficiency)

FY 2007 Compliance with Laws and Regulations

- Substantial Noncompliance:
  - Information Systems Security Controls

OIG Oversight of R. Navarro & Associates, Inc. Performance

To fulfill our responsibilities under the CFO Act and related legislation for ensuring the quality of the audit work performed, we monitored R. Navarro & Associates' audit of NRC's FYs 2007 and 2006 financial statements by:

- Reviewing their approach and planning of the audit,
- Evaluating the qualifications and independence of its auditors,
- Monitoring the progress of the audit at key points,
- Examining the working papers related to planning and performing the audit and assessing NRC's internal control,
- Reviewing R. Navarro & Associates' audit reports to ensure compliance with Government Auditing Standards and OMB Bulletin No. 07-04,
- Coordinating the issuance of the audit reports, and
- Performing other procedures that we deemed necessary.

R. Navarro & Associates, Inc. is responsible for the attached auditors' reports, dated November 7, 2007, and the conclusions expressed therein. The Office of the Inspector General (OIG) is responsible for technical and administrative oversight regarding the firm's performance under the terms of the contract. Our review, as differentiated from an audit in conformance with Government Auditing Standards, was not intended to enable us to express, and accordingly we do not express, an opinion on:

- NRC's financial statements,
- The effectiveness of NRC's internal control over financial reporting, or
- NRC's compliance with laws and regulations.

However, our monitoring review, as described above, disclosed no instances where R. Navarro & Associates, Inc.

did not comply with applicable auditing standards.

Performance Reporting

As required by OMB Bulletin No. 07-04, with respect to internal control related to performance measures determined by management to be key and reported in the Management's Discussion and Analysis, we:

- Obtained an understanding of the design of significant internal controls relating to the existence and completeness assertions, and
- Determined whether they have been placed in operation.

Our procedures were not designed to provide assurance on internal control over performance measures and, accordingly, we do not provide an opinion thereon.

Meeting with the Chief Financial Officer

At the exit conference on November 7, 2007, representatives of the Office of the Chief Financial Officer, OIG, and R. Navarro & Associates, Inc. discussed the issues in the report related to the results of the audit.

Comments of the Chief Financial Officer

In his response, the CFO agreed with the auditors' recommendations. We will follow-up on the CFO's implementation of planned corrective actions during FY 2008. The full text of the CFO's response follows this report.

We appreciate NRC staff's cooperation and continued interest in improving financial management within NRC.

INDEPENDENT AUDITORS' REPORT ON THE FINANCIAL STATEMENTS

2831 Camino Del Rio South, Suite 306
San Diego, California 92108
(619) 298-8193

Chairman Dale E. Klein
U.S. Nuclear Regulatory Commission
Washington, DC

In our audits of the U.S. Nuclear Regulatory Commission (NRC), we found:

- the financial statements are presented fairly, in all material respects, in conformity with accounting principles generally accepted in the United States of America;

- NRC had effective internal control over financial reporting except for the effects of a material weakness related to information systems security controls;

- no instances of noncompliance with laws and regulations, exclusive of the Federal Financial Management Improvement Act (FFMIA), that are required to be reported under applicable audit standards; and

- one instance of substantial noncompliance with the requirements of FFMIA related to information systems security controls.

The following sections provide additional detail about our conclusions and the scope of our audits.

We have audited the accompanying balance sheets of NRC as of September 30, 2007, and 2006, and the related statements of net cost, statements of changes in net position, and statements of budgetary resources for the fiscal years then ended. These financial statements are the responsibility of NRC's management. Our responsibility is to express an opinion on these financial statements based on our audits.

We conducted our audits in accordance with auditing standards generally accepted in the United States of America, the standards applicable to financial audits contained in Government Auditing Standards, issued by the Comptroller General of the United States, and OMB Bulletin No. 07-04, Audit Requirements for Federal Financial Statements. Those standards and the Bulletin require that we plan and perform the audits to obtain reasonable assurance about whether the financial statements are free of material misstatement. An audit includes examining, on a test basis, evidence supporting the amounts and disclosures in the financial statements. An audit also includes assessing the accounting principles used and significant estimates made by management, as well as evaluating the overall financial statement presentation. We believe that our audits provide a reasonable basis for our opinion.

Matters of Emphasis

Classification of Costs

OMB Circular A-136, Financial Reporting Requirements, provides guidance to Federal agencies for presenting program costs classified by intragovernmental and public components. The basis for classification relies on the concept of who received the benefits of the costs incurred (i.e., private sector licensees versus Federal licensees) rather than who was paid. However, following the advice of OMB, NRC classified the costs on the Statements of Net Cost using an underlying concept of who was paid. Furthermore, OMB Circular A-136 requires that the Statement of Net Cost be presented using full program costs by output. NRC presents its costs aggregated by strategic plan programs.

U.S. Department of Energy Expenses

NRC's principal statements include reimbursable expenses of the U.S. Department of Energy (DOE) National Laboratories. For the fiscal years ended September 30, 2007, and 2006, NRC's Statements of Net Cost include approximately $64.4 and $67.8 million, respectively, of reimbursed expenses. Our audits included testing these expenses for compliance with laws and regulations applicable to NRC. The work placed with DOE is under the auspices of a Memorandum of Understanding between NRC and DOE. The examination of DOE National Laboratories for compliance with laws and regulations is DOE's responsibility. This responsibility was further clarified by a memorandum of the Government Accountability Office's (GAO) Assistant General Counsel, dated March 6, 1995, where he opined that "...DOE's inability to assure that its contractors' costs [National Laboratories] are legal and proper...does not compel a conclusion that NRC has failed to comply with laws and regulations." DOE also has the cognizant responsibility to assure audit resolution and provide the results of its audits to NRC.

Opinion

In our opinion, the financial statements referred to above and included in NRC's Performance and Accountability Report present fairly, in all material respects, the financial position of NRC as of September 30, 2007, and 2006, and its net costs, changes in net position, and budgetary resources for the fiscal years then ended in conformity with accounting principles generally accepted in the United States of America.

As discussed in note 14 to the financial statements, NRC changed its presentation of allocation transfers as required by OMB Circular A-136.

Independent Auditors' Report on the Effectiveness of Internal Control Over Financial Reporting

We have examined the effectiveness of NRC's internal control over financial reporting, as of September 30, 2007, based on the criteria in OMB Bulletin No. 07-04. The Bulletin states management is required to establish internal accounting and administrative controls to provide reasonable assurance that transactions are properly recorded, processed, and summarized to permit the preparation of the financial statements in accordance with accounting principles generally accepted in the United States of America and that assets be safeguarded against loss from unauthorized acquisition, use or disposal. NRC's management is responsible for maintaining effective internal control over financial reporting. Our responsibility is to express an opinion on the effectiveness of internal control over financial reporting based on our examination.

Our examination was conducted in accordance with the attestation standards established by the American Institute of Certified Public Accountants (AICPA); the standards applicable to financial statement audits contained in Government Auditing Standards, issued by the Comptroller General of the United States; and OMB Bulletin No. 07-04. Accordingly, we obtained an understanding of the internal control over financial reporting, tested and evaluated the design and operating effectiveness of internal control, and performed such other procedures as we considered necessary in the circumstances. We believe that our examination provides a reasonable basis for our opinion.

Because of inherent limitations in any internal control, misstatements due to error or fraud may occur and not be detected. Also, projections of any evaluation of internal control over financial reporting to future periods are subject to the risk that the internal control may become inadequate because of changes in conditions, or that the degree of compliance with policies or procedures may deteriorate.

We noted certain matters involving the internal control and its operation that we consider to be significant deficiencies under standards established by the AICPA and OMB Bulletin No. 07-04.

A significant deficiency is a deficiency in internal control, or combination of deficiencies, that adversely affects an entity's ability to initiate, authorize, record, process, or report financial data reliably in accordance with generally accepted accounting principles such that there is more than a remote likelihood that a misstatement of an entity's financial statements that is more than inconsequential will not be prevented or detected. As discussed further in this report, the significant deficiencies are related to: (1) weaknesses in NRC's information systems security controls, and (2) weaknesses in NRC's fee billing system.

As defined by OMB Bulletin No. 07-04, a material weakness is a significant deficiency, or combination of significant deficiencies, that results in more than a remote likelihood that a material misstatement of the financial statements will not be prevented or detected. The significant deficiency related to information systems security controls is considered to be a material weakness.

In our opinion, except for the effect of the material weakness described below, NRC has maintained, in all material respects, effective internal control over financial reporting as of September 30, 2007, based on the internal control objectives listed in OMB Bulletin No. 07-04.

Information Systems Security Controls

The Federal Information Security Management Act (FISMA) independent evaluations for fiscal years 2005 and 2006 identified several weaknesses in NRC's information systems security program. The fiscal year 2007 evaluation (Report OIG-07-A-19) identified similar results, including 14 weaknesses of which the following two were considered to be significant deficiencies:

- "Only 2 of the 30 operational NRC information systems have a current certification and accreditation, and only 4 of 11 systems used or operated by a contractor or other organization on behalf of NRC have a current certification and accreditation…. [and]
- Annual contingency testing is still not being performed for all systems."

Based on a self-evaluation of management controls over information systems, NRC concluded that the two significant deficiencies identified in the FISMA report should be reported as a material weakness in its annual statement of assurance required by the Federal Managers' Financial Integrity Act (FMFIA).

Certification and Accreditation

The primary financial management systems consist of three NRC operated systems and two systems operated by the Department of Interior's National Business Center. Of these five systems, three had current certifications and accreditations as of September 30, 2007. The other two operated under interim authorities to operate (IATO) during the entire year. An IATO is a limited authorization to operate an information system under specific terms and conditions and acknowledges greater agency-level risk for a limited period of time. An information system is not considered accredited during the period of limited authority to operate.

Furthermore, although some of the financial management systems have current certifications and accreditations, all financial management systems were and are at risk because they either reside on or rely on a general support system (GSS) which does not have a current certification and accreditation. Therefore, management does not know whether the security controls for the general support systems are adequate, thereby creating potential risk. All NRC information systems that depend on the security controls provided by the general support systems inherit that potential risk.

OMB Circular A-130, Management of Federal Resources, Appendix III, characterizes the absence of authorization to process (certification and accreditation) as an example of a significant deficiency.

Annual Contingency Testing

During fiscal year 2007, annual contingency plan testing was performed for all of the primary financial management systems, however, testing was not performed for 25 of the remaining NRC information systems, including a GSS, and 9 of the remaining contractor operated systems. Contingency plan testing is considered to be a key element of information system security programs, and is essential in determining whether or not plans will function as intended in an emergency situation.

Recommendation

1. The CFO should coordinate with the Office of Information Services and the Executive Director for Operations to ensure that any vulnerabilities of the general support systems and the financial management systems are addressed and resolved timely.

OMB Bulletin 07-04 requires significant deficiencies identified in the FISMA evaluation that are related to financial management systems be reported as a substantial noncompliance with FFMIA. Accordingly, our report on compliance with laws and regulations identifies the FISMA significant deficiencies previously described as a substantial noncompliance with FFMIA, because three systems did not have a current certification and accreditation, and contingency plan testing was not performed for a GSS.

Fee Billing System

NRC is required by law to recover a percentage of its budget authority in each year through fees billed to reactor and materials licensees and applicants. Annual license fees are assessed under 10 CFR Part 171 for nuclear facilities and materials licensees. Other fee types include licensing actions, inspections and other services, established in 10 CFR Part 170. Since fiscal year 2004, we have reported a material weakness related to NRC's fee billing system. The deficiencies reported included: (1) intensive manual processes, (2) the lack of comprehensive quality assurance procedures over the billing process, and (3) the fee billing feeder processes.

The following paragraphs describe the conditions and the agency's progress in addressing them.

- Intensive Manual Processes. Due to the age and design of the Fee Billing System, NRC has evolved an operating style characterized by over-reliance on a small team to prepare, review, and issue billings on a monthly and quarterly basis. The system does not easily provide a drill down capacity to review billing questions.

- Comprehensive Quality Assurance Procedures. NRC's existing quality assurance procedures do not fully address the completeness of billable hours.

- Fee Billing Feeder Processes. In prior years, NRC has identified significant underbillings due to various deficiencies of feeder processes, and a lack of independent checks to validate the completeness of feeder data.

The above conditions continue to affect the fee billing process. However, as discussed below, during fiscal year 2007, NRC took various steps to improve its internal control over the billing cycle.

In February 2007, NRC revised its standard procedures for the Part 170 quarterly certification process used by Headquarters and regional offices. The procedures were revised based on consultations with fee coordinators and an analysis of prior underbilling cases, and were designed to improve the accuracy of the quarterly billings. These procedures were applied to the third and fourth quarter of FY 2007.

NRC also developed a validation application to compare hours recorded in HRMS, the time and labor system, to the hours billed under Part 170. The application was used as a compensating control to validate billings issued for inspections, licensing actions and other billable tasks for the entire fiscal year. This effort resulted in identifying underbillings of approximately $2.6 million that had not been detected by employees in the normal course of performing their assigned billing duties. These results illustrate the need for the continued and expanded application of detection controls to compensate for deficiencies inherent in the current billing process. Further, although the application was used to validate most billings for the entire fiscal year, it was not performed on an ongoing routine basis throughout the entire fiscal year. Finally, we noted that the validation application has not been applied to casework fees.

In fiscal year 2007, we noted that the Part 170 fees receivable of $26 million related to inspections in process at year end were initially understated by $4.7 million because existing controls did not detect that the quarterly "future billables" reports only reflect unbilled hours incurred during a given quarter. At our request, management adjusted the accounts receivable reported in the accompanying financial statements to appropriately reflect all unbilled hours at year end.

While many improvements were made to the fee billing processes in FY 2007, we conclude that the remaining weaknesses constitute a significant deficiency as defined by OMB. However, this finding is no longer considered to be a material weakness.

The GAO's Standards for Internal Control in the Federal Government state, "Internal control should generally be designed to assure that ongoing monitoring occurs in the course of normal operations. It is performed continually and is ingrained in agency's supervisory activities, comparisons, reconciliations, and other actions people take in performing their duties."

Recommendation

2. The CFO should continue to define, design, and implement compensating controls over the fee billing system.

Report on Compliance With Laws and Regulations

We conducted our audit for the year ended September 30, 2007, in accordance with auditing standards generally accepted in the United States of America, the standards applicable to financial audits contained in Government Auditing Standards issued by the Comptroller General of the United States, and OMB Bulletin No. 07-04.

NRC management is responsible for complying with laws and regulations applicable to NRC. As part of obtaining reasonable assurance about whether NRC's financial statements are free of material misstatement, we performed tests of its compliance with certain provisions of applicable regulations, noncompliance with which could have a direct and material effect on the determination of financial statement amounts, and certain other laws and regulations specified in OMB Bulletin No. 07-04, including FFMIA requirements. We limited our tests of compliance to those provisions and we did not test compliance with all laws and regulations applicable to NRC.

The objective of our audit of the financial statements was not to provide an opinion on overall compliance with such provisions of laws and regulations and, accordingly, we do not express such an opinion.

U.S. Department of Energy Expenses

NRC's principal statements include reimbursable expenses of the U.S. Department of Energy (DOE) National Laboratories. For the fiscal years ended September 30, 2007, and 2006, NRC's Statements of Net Cost include approximately $64.4 and $67.8 million, respectively, of reimbursed expenses. Our audits included testing these expenses for compliance with laws and regulations applicable to NRC. The work placed with DOE is under the auspices of a Memorandum of Understanding between NRC and DOE. The examination of DOE National Laboratories for compliance with laws and regulations is DOE's responsibility. This responsibility was further clarified by a memorandum of the GAO's Assistant General Counsel, dated March 6, 1995, where he opined that "...DOE's inability to assure that its contractors' costs [National Laboratories] are legal and proper...does not compel a conclusion that NRC has failed to comply with laws and regulations." DOE also has the cognizant responsibility to assure audit resolution and should provide the results of its audits to NRC.

The results of our tests of compliance with laws and regulations, exclusive of those referred to for FFMIA, disclosed no instances of noncompliance with laws and regulations that are required to be reported under Government Auditing Standards or OMB Bulletin No. 07-04.

Under FFMIA, we are required to report whether NRC's financial management systems substantially comply with the Federal financial management systems requirements, applicable Federal accounting standards, and the United States Government Standard General Ledger at the transaction level. To meet this requirement, we performed tests of compliance with the provisions of FFMIA section 803(a). The results of our tests disclosed one instance, noted below, where NRC's financial management systems did not substantially comply with Federal financial management systems requirements.

In our Report on the Effectiveness of Internal Control Over Financial Reporting, we identified a material weakness in NRC's information systems security controls. We believe that this matter represents substantial noncompliance with the Federal financial management system requirements under FFMIA.

Status of Prior Year Comments

In our prior year report on internal control, we discussed the presence of material weaknesses related to the fee billing system and information systems security controls. During fiscal year 2007, NRC improved its internal control over fee billings by implementing additional detection controls. Consequently, we conclude that the remaining weaknesses in the fee billing system constitute a significant deficiency, as defined by OMB, but the significant deficiency is not considered to be a material weakness. The material weakness related to the information systems security controls continued to exist during fiscal year 2007.

In our prior year report on compliance with laws and regulations we reported a noncompliance related to the development of Part 170 fees and a substantial noncompliance with FFMIA related to the Fee Billing System. Corrective actions have been implemented by NRC to remediate the Part 170 fees noncompliance and that prior finding is now closed. Furthermore, due to certain compensating controls implemented by management in FY 2007, we determined that the remaining weaknesses in the Fee Billing System do not represent substantial noncompliance with FFMIA.

Internal Control Related to Performance Measures

With respect to internal controls related to performance measures described in Chapter 2 of the performance and accountability report, the OIG performed those procedures and will address this issue separately. Our procedures were not designed to provide assurance over reported performance measures and, accordingly, we do not provide an opinion on such information.

Consistency of Other Information

Our audit was conducted for the purpose of forming an opinion on the financial statements of NRC taken as a whole. The required supplementary information referred to as the Management Discussion and Analysis, Chapter 1 of this Performance and Accountability Report, is not a required part of the financial statements but is supplementary information required by OMB Circular A-136. We have applied certain limited procedures which consisted principally of inquiries of management regarding the methods of measurement and presentation of the supplementary information. However, we did not audit the information and express no opinion on it.

The other accompanying information included in Chapter 2 and the appendices to the Performance and Accountability Report, is required by OMB Circular A-136 and is presented for purposes of additional analysis and is not a required part of the financial statements. Such information has not been subjected to the auditing procedures applied in the audit of the financial statements and, accordingly, we express no opinion on it.

Our audit was conducted for the purpose of forming an opinion on the financial statements of NRC taken as a whole. The required supplementary information, Schedule of Budgetary Resources, included in the Performance and Accountability Report, is not a required part of the financial statements but is supplementary information required by OMB Circular A-136. This information is also presented for purposes of additional analysis. This information has been subjected to the auditing procedures applied in the audit of the financial statements and, in our opinion, is fairly stated in all material respects in relation to the financial statements taken as a whole.

We noted certain additional matters that we will report to NRC management in a separate letter.

This report is intended solely for the information and use of NRC management, the Inspector General, OMB, GAO, and the Congress and is not intended to be and should not be used by anyone other than these specified parties.

R. Navarro & Associates, Inc.

November 7, 2007

MANAGEMENT'S RESPONSE TO THE INDEPENDENT AUDITORS' REPORT ON THE FINANCIAL STATEMENTS

UNITED STATES
NUCLEAR REGULATORY COMMISSION
WASHINGTON, D.C. 20555-0001

November 9, 2007

OFFICE OF THE
CHIEF FINANCIAL OFFICER

MEMORANDUM TO: Stephen D. Dingbaum
Assistant Inspector General for Audits
Office of the Inspector General

FROM: William M. McCabe
Chief Financial Officer

SUBJECT: AUDIT OF THE FISCAL YEARS 2007 AND 2006 FINANCIAL STATEMENT AUDITS

We appreciate the collaborative relationship between the Office of the Inspector General, the auditors and the Office of the Chief Financial Officer in supporting our continuing effort to improve financial reporting. We have reviewed the independent auditors' report of the Agency's Fiscal Year 2007 and 2006 financial statements and are in general agreement with the report and overall findings.

Our responses to the recommendations follow:

Recommendation 1

The Chief Financial Officer (CFO) should coordinate with the Office of Information Services and the Executive Director for Operations to ensure that any vulnerabilities of the general support systems and the financial management systems are addressed and resolved timely.

Response

Agree. The CFO will continue to coordinate with the Office of Information Services and the Executive Director for Operations to ensure that vulnerabilities of the general support systems and the financial management systems are addressed timely.

Recommendation 2

The CFO should continue to define, design, and implement compensating controls over the fee billing system.

Response

Agree. The CFO will continue to assess and implement opportunities to improve the internal controls over the fee billing system.

Appendices

Refueling a Nuclear Reactor

The Oconee nuclear station is located in Seneca, SC.

INSPECTOR GENERAL'S ASSESSMENT OF THE MOST SERIOUS MANAGEMENT AND PERFORMANCE CHALLENGES FACING THE NRC

OFFICE OF THE
INSPECTOR GENERAL   September 28, 2007

MEMORANDUM TO: Chairman Klein

FROM: Hubert T. Bell   *Hubert T. Bell*
             Inspector General

SUBJECT: INSPECTOR GENERAL'S ASSESSMENT OF THE MOST SERIOUS
             MANAGEMENT AND PERFORMANCE CHALLENGES FACING
             THE NUCLEAR REGULATORY COMMISSION (OIG-07-A-20)

The Reports Consolidation Act of 2000 requires the Inspector General of each Federal agency to annually summarize what he or she considers to be the most serious management and performance challenges facing the agency and to assess the agency's progress in addressing those challenges. In accordance with the Act, I identified eight management and performance challenges that I consider to be the most serious. The list of eight challenges reflects the consolidation of the prior challenges 4 and 9 resulting in the following description for new challenge 4: Ability to modify regulatory processes to meet a changing environment, specifically the potential for a nuclear renaissance.

We appreciate the cooperation extended to us during this evaluation. The agency provided comments on this report, which have been incorporated, as appropriate. If you have any questions or comments about this report, please feel free to contact Stephen D. Dingbaum, Assistant Inspector General for Audits, at 415-5915 or me at 415-5930.

cc: Commissioner Jaczko
    Commissioner Lyons

TABLE OF CONTENTS FOR APPENDIX A

EXECUTIVE SUMMARY

Background

The Reports Consolidation Act of 2000 (the Act) requires the Inspector General (IG) of each Federal agency to annually summarize what he or she considers to be the most serious management and performance challenges facing the agency and to assess the agency's progress in addressing those challenges.

Purpose

In accordance with the Act, the IG at the Nuclear Regulatory Commission (NRC) updated what he considers to be the most serious management and performance challenges facing NRC. As part of the evaluation, the Office of the Inspector General staff sought input from NRC's Chairman, Commissioners, and NRC management to obtain their views on what challenges the agency is facing and what efforts the agency has taken to address previously identified management challenges.

Results In Brief

The IG identified eight challenges that he considers are the most serious management and performance challenges facing NRC. The challenges identified represent critical areas or difficult tasks that warrant high-level management attention.

In addressing this year's challenges we combined the prior challenge number 4, Ability to modify regulatory processes to meet a changing environment and the prior challenge number 9, Ability to meet the demand for licensing new reactors. The consolidation of these challenges resulted in the following description for new challenge 4: Ability to modify regulatory processes to meet a changing environment, specifically the potential for a nuclear renaissance. We combined the two challenges because the anticipated workload associated with preparing to receive and then review new reactor license applications will strain the agency's current resources and intensify other challenges in NRC's regulatory environment.

The chart that follows provides an overview of the eight most serious management and performance challenges as of September 28, 2007.

MOST SERIOUS MANAGEMENT AND PERFORMANCE CHALLENGES FACING THE NUCLEAR REGULATORY COMMISSION*
AS OF SEPTEMBER 28, 2007
(AS IDENTIFIED BY THE INSPECTOR GENERAL)

Challenge 1 Protection of nuclear material used for civilian purposes.

Challenge 2 Appropriate handling of information.

Challenge 3 Development and implementation of a risk-informed and performance-based regulatory approach.

Challenge 4 Ability to modify regulatory processes to meet a changing environment, specifically the potential for a nuclear renaissance.

Challenge 5 Implementation of information technology.

Challenge 6 Administration of all aspects of financial management.

Challenge 7 Communication with external stakeholders throughout NRC regulatory activities.

Challenge 8 Managing human capital.

* The most serious management and performance challenges are not ranked in any order of importance.

Conclusion

The eight challenges contained in this report are distinct, yet are interdependent to accomplishing NRC's mission. For example, the challenge of managing human capital affects all other management and performance challenges.

The agency's continued progress in taking actions to address the challenges presented should facilitate successfully achieving the agency's mission and goals.

I. BACKGROUND

On January 24, 2000, Congress enacted the Reports Consolidation Act of 2000, requiring Federal agencies to provide financial and performance management information in a more meaningful and useful format for Congress, the President, and the public. The Act requires the Inspector General (IG) of each Federal agency to annually summarize what he or she considers to be the most serious management and performance challenges facing the agency and to assess the agency's progress in addressing those challenges.

II. PURPOSE

In accordance with the Act's provisions, the IG at the NRC updated what he considers to be the most serious management and performance challenges facing NRC. The IG evaluated the overall work of the Office of the Inspector General (OIG), the OIG staff's general knowledge of agency operations, and other relevant information to develop and update his list of management and performance challenges.

In addition, OIG sought input from NRC's Chairman, Commissioners, management and staff to obtain their views on what challenges the agency is facing and what current and future efforts the agency has taken to address previously identified management and performance challenges.

III. EVALUATION RESULTS

The NRC's mission is to "License and regulate the Nation's civilian use of byproduct, source, and special nuclear materials to ensure adequate protection of public health and safety, promote the common defense and security, and protect the environment." Like other Federal agencies, NRC faces management and performance challenges in carrying out its mission.

Determination of Management and Performance Challenges

Congress left the determination and threshold of what constitutes a most serious management and performance challenge to the discretion of the Inspectors General. As a result, the IG applied the following definition in identifying challenges:

> **Serious management and performance challenges are mission critical areas or programs that have the potential for a perennial weakness or vulnerability that, without substantial management attention, would seriously impact agency operations or strategic goals.**

Based on this definition, the IG revised his list of the most serious management and performance challenges facing NRC. The challenges identified represent critical areas or difficult tasks that warrant high-level management attention. The following chart provides an overview of the eight management challenges. The sections that follow the chart provide more detailed descriptions of the challenges, descriptive examples related to the challenges, and examples of efforts the agency has taken or are underway to address the challenges.

Changes to Management Challenges

This year's challenges are essentially the same as last year, with two exceptions.

Description Change - Challenge 2

Last year's challenge 2: *Protection of information* was changed this year to *Appropriate handling of information*. The focus has been broadened to include emphasis on the importance of releasing information that the public has a right to know while protecting sensitive information that should not be released.

Integrating Challenges 4 and 9

Last year's challenge 9[1] and challenge number 4[2] were combined this year to form challenge number 4 which reads, *Ability to modify regulatory processes to meet a changing environment, specifically the potential for a nuclear renaissance.*

The prior two challenges were combined because the anticipated workload associated with preparing to receive and then review new reactor license applications will strain the agency's current resources and intensify other challenges in NRC's regulatory environment. While responding to the emerging demands associated with regulating new reactors, NRC must also sustain technical quality in carrying out its current regulatory responsibilities.

CHALLENGE 1

Protection of nuclear material used for civilian purposes.

NRC grants licenses for the possession and use of radioactive materials and establishes regulations to govern the possession and use of those materials. NRC's regulations require that certain materials licensees have extensive material control and accounting programs as a condition of their licenses. All other licensees (including those requesting authorization to possess small quantities of special nuclear materials) must develop and implement plans that demonstrate a commitment to accurately control and account for radioactive materials.

The issues related to this challenge and the agency's actions to address each issue include the following:

Issue: Ensure that radioactive material is adequately protected to preclude it from being used for malicious purposes.

Action: NRC is enhancing its materials licensing processes, which include a new policy that requires on-site visits before NRC issues new material licenses; is examining existing licenses to determine their legitimacy; and is forming a working group to update and revise existing materials guidance.

Issue: Develop and implement a system to ensure the accurate tracking of byproduct material, especially those materials with the greatest potential to impact public health and safety.

Action: NRC has published its final rulemaking on the National Source Tracking System (NSTS) and is working to develop and implement systems [NSTS and Web-based Licensing] for tracking materials and licenses.

Issue: Ensure reliable control and accounting of special nuclear materials in the NRC and Department of Energy's (DOE) jointly managed Nuclear Materials Management and Safeguards System (NMMSS).

Action: NRC has taken steps to ensure that licensees comply with material control and accounting (MC&A) requirements. For example, revisions to Inspection Manual Chapter 2800 and Temporary Instruction 2515/154 required that NMMSS book balances be compared to actual inventories possessed by reactor licensees as well as licensees holding small amounts of special nuclear materials.

Issue: Provide adequate inspection to verify the control and accountability of special nuclear materials at licensee sites.

Action: The staff proposed an MC&A rulemaking plan early in 2007 that will enhance MC&A regulations, inspections, and licensing. Among the enhancements are requirements

---

[1] 2006 Management Challenge 9: *Ability to meet the demand for licensing new reactors.*
[2] 2006 Management Challenge 4: *Ability to modify regulatory processes to meet a changing environment.*

to conduct periodic inspections to verify that material licensees comply with MC&A requirements.

CHALLENGE 2

Appropriate handling of information.

NRC is required to appropriately protect and withhold information from public disclosure for reasons of security, personal privacy, or commercial or trade secrets protection. The agency also has a duty to release information the public has a right to know. NRC's goal is to strike an appropriate balance between a regulatory process that is open to the public and the protection from disclosure of sensitive information, which would be useful to potential adversaries. NRC traditionally has given the public access to a significant amount of information about the facilities and materials the agency regulates.[3] The Atomic Energy Act, subsequent legislation, and various NRC regulations have given the public the right to participate in the licensing and oversight process for NRC licensees.

The issues related to this challenge and the agency's actions to address each issue include the following:

Issue: Ensure that information is released to the public that the public has a right to know.

Action: After receiving congressional criticism, NRC gave the public access to documents associated with a uranium spill that had been previously designated as Official Use Only (not releasable to the public). Further, the Commission is reconsidering its policy and criteria for withholding information from the public.

Issue: Appropriately protect and withhold information from public disclosure, especially information related to personally identifiable information (PII), security related information and safeguards information (SGI).

Action: NRC has conducted searches and promptly removed all documents containing PII from public availability after inadvertent disclosure. In addition, NRC has established the PII Task Force to identify how PII is used and to develop policies and procedures to protect this information while minimizing the impact on agency operations. NRC's PII Task Force also developed a draft breach notification policy as required by the Office of Management and Budget.

Action: NRC issued SGI Fingerprinting Orders that require any person who seeks or obtains access to SGI to undergo a Federal Bureau of Investigation identification and criminal history check based on that individual's fingerprints.

CHALLENGE 3

Development and implementation of a risk-informed and performance-based regulatory approach.

NRC must increase its safety and security focus on licensing and oversight activities through the application of a balanced combination of experience, deterministic models, and probabilistic analysis. This approach is known as risk-informed and performance-based regulation. Incorporating risk analysis into regulatory decisions is intended to improve the regulatory process by focusing NRC and licensee attention and actions on the highest risk areas.

---

[3] Openness has been and remains a cornerstone of NRC's regulatory philosophy.

The issues related to this challenge and the agency's actions to address each issue include the following:

Issue: Ensure that the appropriate level of focus on risk-informed and performance-based regulation is maintained.

Action: NRC is continuing its work to improve the agency's Risk-Informed and Performance-Based Plan,[4] including a recent expansion of the Plan's objectives to more fully achieve a risk-informed and performance-based regulatory structure.

Issue: Develop and implement risk-informed and performance-based regulation for fuel cycle facilities.

Action: NRC is preparing a framework for the fuel cycle facility oversight program.

Issue: Ensure that the Reactor Oversight Process meets the agency's regulatory needs.

Action: NRC uses results of an annual self-assessment of the Reactor Oversight Process to better identify significant performance issues and to ensure that licensees take appropriate actions to maintain acceptable safety and security performance.

Issue: Ensure that research programs enhance the validity of current risk models, and also develop risk insights for new technologies, including program areas now transitioning to risk-informed regulation (e.g., fire protection).

Action: NRC is developing and implementing a formal written process for maintaining probabilistic risk assessment models that are sufficiently representative of the as-built, as-operated plants to support model uses.

CHALLENGE 4

Ability to modify regulatory processes to meet a changing environment, specifically the potential for a nuclear renaissance.

While NRC maintains its core regulatory programs, it must adapt to emerging changes in the regulatory environment. Specifically, the agency must maintain the rigor of its regulation of the current fleet of operating reactors while simultaneously preparing for an influx of applications for new reactors. Furthermore, the agency must be ready to regulate facilities using new fuel processing technologies and address issues relating to the disposal of increasing quantities of radioactive waste.

The issues related to this challenge and the agency's actions to address each issue include the following:

Issue: Maintain the ability to review operating reactor licensee applications for license renewals and power uprates submitted by industry in response to the Nation's demand for energy production.

Action: NRC is continuing its work with operating reactor plant licensees to develop a schedule of anticipated license amendment requests for license renewals and power uprates.

Issue: Develop and create the infrastructure necessary to support the review of new plant licensing applications, to include: reinstituting the Construction Inspection Oversight program, developing strong control processes for project management to ensure the agency meets its new reactor review and licensing objectives, developing technical review processes and ensuring that NRC implements a comprehensive standard review plan and adequately documented safety evaluation reports.

---

[4] The Risk-Informed and Performance-Based Plan was formerly known as the Risk-Informed Regulation Implementation Plan.

Action: NRC is preparing for the expected receipt of utility applications for new reactor licenses. NRC is issuing reactor design certifications, revising the regulation that governs early site permits, and engaging in ongoing interactions with plant vendors and utilities regarding prospective new reactor applications and licensing activities.

Issue: Ensure that Agreement State programs are adequate and compatible with NRC's program to protect public health and safety and the environment.

Action: NRC continues to conduct about 10-12 reviews per year of Agreement State radioactive materials programs under NRC's Integrated Materials Performance Evaluation Program.

Issue: Address increasing quantities of radioactive waste requiring interim or permanent disposal sites.

Action: NRC has conducted a review for dry cask waste storage systems.

Action: NRC is currently assessing its overall low-level waste program to prioritize ongoing and future staff actions and activities, along with associated schedules and resource estimates.

Issue: Prepare for and respond to delays and uncertainties related to its receipt and review of a DOE license application to construct a high-level radioactive waste repository at Yucca Mountain.

Action: NRC continues to prepare for receipt of DOE's license application to construct a high-level waste repository, which is expected in July 2008. NRC is focused on pre-licensing activities, issuing interim staff guidance, identifying the application review approach, and identifying review teams

---

CHALLENGE 5

Implementation of information technology.

NRC needs to upgrade and modernize its information technology (IT) capabilities both for employees and for public access to the regulatory process. Recognizing the need to modernize, the Office of Information Services established goals to improve the productivity, efficiency, and effectiveness of agency programs and operations, and enhance the use of information for all users inside and outside the agency.

The issues related to this challenge and the agency's actions to address each issue include the following:

Issue: Ensure that information systems are protected.

Action: NRC has made little progress in correcting the following two significant deficiencies concerning its Information System-wide Security Controls. Annual contingency plan testing is not being performed, and only 2 of 30 systems have been assessed to determine risks to agency operations, agency assets, or individuals, resulting in a failing grade from Congress for computer security. Although the agency is working towards certification and accreditation for all of its systems, the agency does not expect to accomplish this goal until the end of FY 2009. Actions also include awarding a multimillion dollar contract to enhance agencywide information systems security, documenting the process to complete certifications and accreditations of systems and categorizing systems as to sensitivity of the information.

Issue: Upgrade and manage IT activities to improve the productivity, efficiency, and effectiveness of agency programs and operations.

Action: NRC recognizes that it lags behind many other Federal agencies in terms of its IT infrastructure. For example, the ability to support technologies such as wireless and Microsoft Office suite, which is already the standard software used in the private sector and much of the public sector. In addition, the agency is evaluating options for replacing its aging applications such as the Agencywide Documents Access and Management System and Human Resources Management System. The agency has developed an information technology/information management strategic plan that addresses infrastructure planning and seeks a single, integrated infrastructure technology roadmap as part of an overall enterprise architecture transition plan.

Issue: Maintain a knowledgeable information technology staff.

Action: NRC is continuing to upgrade its IT infrastructure to a state of the art level, therefore, NRC must hire and retain staff who possess the required expertise. NRC has initiated new workforce planning strategies to address this, to include offering higher pay grades/salaries for needed proficiencies, keeping vacancy announcements indefinitely open to fill the many vacancies, and paying relocation expenses.

CHALLENGE 6

Administration of all aspects of financial management.

NRC management is responsible for establishing and maintaining effective internal controls and financial management systems that meet the objectives of several statutes including the Federal Managers' Financial Integrity Act. This Act mandates that NRC establish controls that reasonably ensure that (1) obligations and costs comply with applicable law; (2) assets are safeguarded against waste, loss, unauthorized use, or misappropriation; and (3) revenues and expenditures are properly recorded and accounted for. This Act encompasses program operational, and administrative areas, as well as accounting and financial management.

The issues related to this challenge and the agency's actions to address each issue include the following:

Issue: Resolve the material weaknesses[5] reported in the audit of NRC's financial statements and the issues related to licensee fee reporting.

Action: NRC is addressing the continuing material weakness by assessing all processes and system interfaces associated with the fee billing process and system to ensure controls are adequate. The agency has implemented a number of new and improved controls including a validation tool which analyzes and reconciles the completeness and accuracy of billing for reactors and materials inspections. As a result, the agency has decreased the risk of potential billing errors and further enhanced the control environment.

Action: NRC conducted a business process improvement study focused on time and labor and fee billing processes. The study made a number of recommendations for improvement including the need to corporately manage the reporting codes and to reduce the number of reporting codes to improve internal controls. As a result, the agency has developed interim guidance for managing reporting codes and expects to reduce the number of codes (currently totaling approximately 9,500) by another 1,000 before year-end.

---

[5] FY 2006 financial statement internal control reportable conditions include a continuing material weakness regarding the Fee Billing System and a new material weakness regarding lack of required Information System-wide Security Controls. Discussion of the actions taken concerning the latter reportable condition is contained in Challenge 5 – Implementation of information technology.

Issue: Replace NRC's current financial systems which are obsolete, overly complex, and inefficient.

Action: NRC has submitted a business case that recommended the replacement of five aging financial systems[6] with a single integrated core financial system, expected to be operational in October 2009.

CHALLENGE 7

Communication with external stakeholders throughout NRC regulatory activities.

The NRC has stated that nuclear regulation is the public's business and, therefore, it should be transacted in an open and candid manner in order to maintain the public's confidence. The continuing challenge for management is to ensure that there are effective ways of communicating with external stakeholders. Effective communication is vital to the agency's ability to achieve its goals, to include enhancing the public's confidence in NRC's effectiveness as a regulator.

The issues related to this challenge and the agency's actions to address each issue include the following:

Issue: Ensure effective interaction with a diverse group of external stakeholders (e.g., industry, Congress, general public, other Federal agencies, citizen groups) by providing clear, accurate, and timely information about NRC's regulatory activities.

Action: NRC provides a quarterly report on the status of its licensing and other regulatory activities to the Senate Subcommittee on Clean Air and Nuclear Safety.

Action: NRC continues to hold public meetings throughout the year, as well as, an annual public Regulatory Information Conference on specific licensing and regulatory activities to share information with stakeholders.

Issue: Ensure compliance with the Freedom of Information Act (FOIA) regarding disclosure of information to the public, through both FOIA requests and FOIA automatic disclosure requirements, and timely responses to FOIA requests.

Action: NRC is implementing revised Internal Commission Procedures to require a review of Commission decision documents to determine whether these documents should be released, in whole or in part, in accordance with the automatic disclosure provisions of FOIA.

CHALLENGE 8

Managing human capital.

NRC's human capital needs will undergo changes due to the expected receipt of (1) applications to construct and operate the next-generation of nuclear reactors, (2) DOE's license application for a nuclear waste repository, and (3) industry applications to increase the number of fuel cycle facilities. By FY 2009, NRC will have hired approximately 1,200 new employees. Moreover, a United States Government Accountability Office report issued January 2007,[7] found that about 16 percent of NRC employees are eligible to retire, a figure that is expected grow to 33 percent by FY 2010.

The issues related to this challenge and the agency's actions to address each issue include the following:

---

[6] The five financial systems are Federal Financial System, Fee Billing System, Allotment/Allowance Financial Plan System, Cost Accounting System, and the Capitalized Property System.

[7] *Human Capital: Retirements and Anticipated New Reactor Applications Will Challenge NRC's Workforce*, GAO-07-105, January 17, 2007.

Issue: Addressing anticipated increased workload demands and retirements.

Action: NRC is recruiting a skilled workforce that targets the anticipated changes facing the agency.[8] The agency is on track to exceed its FY 2007 hiring goal of a net gain of approximately 200 staff.

Action: NRC is enhancing its reactor technology curriculum to meet the demand of its increased and varied workload which includes the review and licensing of the new generation of commercial nuclear reactors.

Action: NRC is implementing knowledge management[9] strategies that include mentoring; early replacement hiring; rehiring annuitants with or without use of a pension offset as applicable[10]; and developing a knowledge management Web site, expressly for the purpose of retaining knowledge before key employees are promoted or retire.

Action: NRC is working with the General Services Administration to acquire additional off-site office space near its headquarters, for up to 300 staff by the late summer of 2008. Furthermore, most NRC regional offices are seeking new office space for additional staff in order to meet increased workload demands.

IV. CONCLUSION

The eight challenges contained in this report are distinct, yet are interdependent to accomplishing NRC's mission. For example, the challenge of managing human capital affects all other management and performance challenges.

The agency's continued progress in taking actions to address the challenges presented should facilitate successfully achieving the agency's mission and goals.

ATTACHMENT A - SCOPE AND METHODOLOGY

This evaluation focused on the IG's annual assessment of the most serious management and performance challenges facing the NRC. The challenges represent critical areas or difficult tasks that warrant high level management attention. To accomplish this work, the OIG focused on determining (1) current challenges, (2) the agency's efforts to address the challenges during FY 2007, and (3) future agency efforts to address the challenges.

The OIG reviewed and analyzed pertinent laws and authoritative guidance. In addition, OIG conducted interviews with agency officials at NRC Headquarters and conducted interviews by telephone with agency officials in the four NRC regional offices. The purpose of the interviews was to identify current performance and management challenges and steps taken by the agency to address these challenges through planning and in daily operations. Since challenges affect mission critical areas or programs that have the potential to impact agency operations or strategic goals, NRC Commission members, the Executive Director for Operations and the Chief Financial Officer were afforded the opportunity to share any information and insights on this subject.

OIG conducted this evaluation from June through August 2007. The major contributors to this report were Steven Zane, Team Leader, Beth Serepca, Team Leader, Sherri Miotla, Team Leader, Vicki Foster, Audit Manager, Michael Steinberg, Senior Auditor, and Lori Konovitz, Senior Analyst.

---

[8] As of the last pay period in July 2007, there were approximately 3,526 NRC staff.

[9] Knowledge management involves capturing critical information and making the right information available to the right people at the right time to assure that knowledge and experience of the current staff is passed on to the next generation of NRC staff.

[10] This flexibility allows NRC to rehire a retiree to fill a position at full pay if the agency has experienced difficulty in filling a position, or if a temporary emergency exists.

NRC ACTIONS RESPONDING TO THE OFFICE OF THE INSPECTOR GENERAL'S MOST SERIOUS MANAGEMENT AND PERFORMANCE CHALLENGES

Below are the NRC's major FY 2007 actions taken in response to the Office of the Inspector General's eight most serious management challenges dated September 28, 2007. While the earlier appendix containing the Office of the Inspector General's management challenges described certain agency activities responding to those challenges, this section represents a more detailed explanation to describe the staff's actions.

CHALLENGE 1

Protection of nuclear material used for civilian purposes.

In FY 2007, the agency issued the proposed rule *Regulatory Improvements to the Nuclear Materials Management and Safeguards System* for public comment. This rule seeks improvement in the accuracy of inventory information for licensees' possession of special nuclear material (SNM) in the Nuclear Materials Management and Safeguards System (NMMSS) database. Currently, licensees possessing 350 grams or more of SNM are required to report their physical inventory results of SNM to the database. This proposed rule would require licensees possessing 1 gram or more of SNM to report their physical inventory to the NMMSS database and to reconcile their physical inventory results with the database. In addition, this proposed rule would require those licensees who have moved SNM into on-site waste type accounts to report and reconcile the quantity of inventory in these holding accounts with the database. If adopted, this proposed rule will help ensure that the NMMSS database contains the most accurate information possible for each licensee. In addition to regulations requiring that certain materials licensees have extensive material control and accounting, substantial work has also been done as well in the reactor safety program.

Additionally, in response to the NRC's request, the Department of Energy (DOE) has tasked the NMMSS operator to promptly process licensee inventory reports that are submitted for entry into the database. DOE has established a performance metric for the NMMSS operator to complete the reconciliation process between NMMSS and the licensee within 30 days of receipt of licensee inventory reports. The status of the reconciliation process for all licensees is documented in the monthly letter status report prepared by the NMMSS operator. Finally, the operator has established a process to notify each licensee regarding whether the licensee inventory reports and NMMSS database are in agreement, and if not, what actions are needed to rectify the inventory.

In May 2007, staff members from the U. S. Government Accountability Office (GAO) notified NRC staff of the results of an investigation where GAO staff was able to obtain a valid NRC radioactive materials license, authorizing the possession of portable gauges containing radioactive sources, using false information (e.g., company name, address, etc.). GAO staff was also able to modify the license using computer software to make it appear to authorize a much greater number of gauges than the original license. NRC immediately suspended the review of all new applications for materials licenses until interim corrective actions were implemented. In response to the GAO's investigation and resulting recommendations, NRC staff developed an action plan to address the efficiency of the interim actions, as well as detail on other longer-term modifications to the NRC's licensing process that would enhance the NRC's and the Agreement States' abilities to verify the validity of license applicants. Multiple groups have been tasked with reviewing the NRC's licensing process and recommending potential improvements. First, the Pre-Licensing Working Group was reconvened to recommend short-term fixes while longer-term solutions are being evaluated by two other groups. An independent, external review panel will analyze the NRC's overall materials security program, including a review of lower-risk sources. The other group to recommend longer-term solutions is the Materials Program Working Group, which will review the efforts of the first two groups and make its own recommendations for improvements in the regulatory process. All three groups involve Agreement State participation.

Orders imposing requirements on transport of radioactive material in quantities of concern (RAMQC) were issued to licensees on July 19, 2005, and licensees were required to implement these orders by January 19, 2006. During FY 2007, the staff surveyed the regulated community to assess the impacts of the RAMQC orders. This survey did not identify any significant adverse issues with implementation of the RAMQC Order. The staff will next move towards incorporating these Orders into the 10 CFR Part 73 regulations.

In FY 2008, the staff will continue the development of the National Source Tracking System (NSTS) and maintenance of the Interim Source Database. Several rulemakings will be initiated in order to expand the NSTS, limit the amount of radioactive material that can be possessed by general licensees, and ensure security for the transportation of radioactive material. The staff also expects to issue a final rule specifying post-9/11 security requirements for a geologic repository operations area (GROA). This rule will amend the applicable NRC regulations to revise the security requirements and material control and accounting (MC&A) requirements for a GROA, and will include new requirements for specific training enhancements, improved access authorization, and enhancements to defensive strategies. The goal of this rulemaking is to ensure that effective security measures are in place for the protection of high-level radioactive waste.

CHALLENGE 2

Appropriate handling of information.

In FY 2007, the NRC established the Personally Identifiable Information (PII) Task Force to identify how PII is used at the NRC and to develop policies

and procedures to protect PII while minimizing the impact on agency operations. The objectives of the task force include: 1) identifying current data sources containing PII; 2) reviewing the use of social security numbers and other PII to reduce the collection and storage of PII; 3) recommending modifications to business processes and operations to protect PII; and 4) increasing staff awareness of PII issues, policies, and procedures. The NRC also created a "PII Project" Web site and maintains a site related to the NRC's Sensitive Unclassified Non-Safeguards Information (SUNSI) program on the NRC's intranet. The Web sites provide NRC staff with current information related to PII and SUNSI activities at the agency as well as links to the NRC's policy for SUNSI and PII.

The NRC completed a review of the agency's shared drives for PII to ensure it is adequately protected or removed, as appropriate. Also, the NRC developed a policy for future periodic reviews of the agency shared drives.

At the March 2007 Regulatory Information Conference, the NRC chaired a session regarding the agency's SUNSI program. This session focused on the four types of SUNSI that most affect external stakeholders who submit documents to the NRC: security-related information, proprietary information, PII, and information under the control of other Federal agencies, state and foreign governments, and international agencies. Emphasis was placed on the importance of protecting PII, the proper way to mark submitted documents, and submitter responsibilities.

The NRC issued Regulatory Issue Summary (RIS) 2007-04, "Personally Identifiable Information Submitted to the U.S. Nuclear Regulatory Commission," to enhance the awareness of permit holders and licensees about PII and the need to protect it from inappropriate disclosure. The RIS is available on the following Web site: http://www.nrc.gov/reading-rm/doc-collections/gen-comm/reg-issues/2007/ri200704.pdf.

In FY 2008, the NRC will establish and implement a plan to eliminate the unnecessary collection and use of Social Security numbers, in response to Office of Management and Budget (OMB) memorandum (M-07-16), "Safeguarding Against and Responding to the Breach of Personally Identifiable Information." In addition, the NRC will determine: 1) whether any NRC contracts require contractors to obtain or possess PII; and 2) if so, whether such possession and use is critical to carry out the contract. Staff will continue to review agency shared drives to ensure that PII is adequately protected or removed. The PII Task Force will continue to identify ways to protect PII and implement changes required by OMB and the Office of Personnel Management (OPM). Finally, staff will continue to participate in the Interagency Best Practices Collaborative meetings sponsored by the Social Security Administration.

In FY 2007, after releasing information regarding a March 2006 spill of high-enriched uranium at a fuel facility, NRC revised its policy to increase the amount of information made publicly available due to security concerns. This policy change also reinforced the public's hearing rights under the Atomic Energy Act, Section 189A. In September 2007, NRC staff issued the Communication Plan for the Release of Redacted Licensing and Enforcement Documents for Fuel Facilities. In preparation for the September 2007 release of redacted documents, the staff reviewed past licensing actions for all license amendments processed from January 1, 2004, to present and identified whether any of the incoming licensee requests were classified documents. A list of licensing actions and security orders that have been redacted and made publicly available is posted on the NRC's public Web site at http://www.nrc.gov/about-nrc/regulatory/adjudicatory/hearing-license-applications.html#7.

CHALLENGE 3

Development and implementation of risk-informed and performance-based regulatory approach.

The agency took significant steps in FY 2007 to enhance communication and implementation of risk-informed and performance-based initiatives.

A Risk-informed Performance-Based Plan (RPP) was implemented to improve the existing approach and transform it into an integrated master plan for activities designed to help the agency achieve the Commission's goal of a holistic, risk-informed and performance-based regulatory structure. The agency has also increased its dialogue with stakeholders via more frequent public meetings to discuss implementation and policy concerns, and to clarify NRC positions.

The agency also made significant progress in the development of human reliability analysis (HRA)-informed products to be used by staff involved with medical applications of byproduct materials. Specifically, based on feedback in previous reviews, the NRC revised HRA-informed training materials that, in combination with an HRA-informed job aid, are intended to help NRC staff to (1) better understand the potential causes of human errors in medical applications of byproduct materials, and (2) use this understanding to justify staff recommendations (e.g., approval of license applications or amendments, acceptance of licensee-proposed corrective actions). In FY 2008, the NRC plans to further develop the HRA-informed job aid, which currently is in prototype form, and to obtain feedback from potential users.

Enhancements to the Generic Issues Program (GIP) have been designed to ensure comprehensive and timely resolution of future generic issues. Implementation will be via a revision to Management Directive 6.4, "Generic Issues Program." The objective is to reserve for GIP review only those issues that have significant generic implications related to risk or security that cannot be more effectively handled by other regulatory programs and processes. The agency plans to employ enhanced risk-informed techniques developed from existing initiatives such as the Accident Sequence Precursor Program to improve the timely assessment of these generic issues.

During FY 2007, Revision 1 to Regulatory Guide 1.200, "An Approach for Determining the Technical Adequacy of Probabilistic Risk Assessment Results for Risk-Informed Activities," was issued. This revision endorses the American Society of Mechanical Engineers (ASME) Probabilistic Risk Assessment (PRA) standard including the appendices addressing large early release frequency and peer review. This regulatory guide describes an acceptable approach for determining that the quality of a licensee's PRA, in total or limited only to those components used to support a licensing action, is expected to instill confidence in the reported PRA results sufficient to allow its use in risk-informed regulatory decision making for light-water reactors. The industry is expected to begin implementing this revised regulatory guide in FY 2008.

The agency completed a PRA of a dry cask storage system at an independent spent fuel storage installation in FY 2007. The study covered various phases of the dry cask storage process, from loading fuel from the spent fuel pool, preparing the cask for storage and transferring it outside the reactor building, to moving the cask to the storage pad and storing it there for 20 years. The study provides risk insights that will be used to further risk-inform license reviews in major technical disciplines and in the update of the Standard Review Plans (SRPs) for dry cask storage and storage facilities. Draft updated SRPs are expected to be issued for public comment in the last quarter of FY 2008.

In FY 2007, the NRC issued a 30-year license to United States Enrichment Corporation (USEC), Inc., to construct and operate a gas centrifuge facility at Piketon, Ohio. The staff performed a risk-informed and performance-based review of the application, allowing a timely review that focused on safety aspects of the facility design.

In FY 2008, the agency will continue working with the American Nuclear Society (ANS) and ASME in the development and endorsement of various PRA quality standards addressing fires, external events, and low power and shutdown operations. In addition, the agency will continue to develop guidance documents to improve the quality of PRAs in support of risk-informed decision-making. One

particularly complex area is the modeling of digital instrumentation and control systems in a PRA. In FY 2008, the agency will continue the development of models and methods for assessing the contribution to risk from digital instrumentation and control systems. Staff will also continue to develop improvements for the Standardized Plant Analysis Risk models used to support the reactor oversight process.

CHALLENGE 4

Ability to modify regulatory processes to meet a changing environment, specifically the potential for nuclear renaissance.

The NRC staff is engaged in numerous ongoing interactions with vendors and utilities regarding prospective new reactor applications and licensing activities. Based on these interactions, the staff expects to receive a significant number of new reactor combined operating license applications over the next several years and is currently developing the infrastructure necessary to support the application reviews.

The NRC has issued design certifications for four reactor designs that can be referenced in an application for a nuclear power plant and is currently performing the design certification review of General Electric's Economic Simplified Boiling Water Reactor design and the Westinghouse AP1000 design. In addition, staff is performing design certification pre-application reviews for AREVA's Evolutionary Power Reactor and Mitsubishi's U.S. Advanced Pressurized Water Reactor. Revised draft standard review plans (SRPs) were issued to design certification applicants in August 2007.

In August 2007, the NRC also issued revisions to the regulation governing early site permits (ESPs), design certifications, and combined licenses (10 CFR Part 52) to improve the effectiveness and efficiency of the licensing process for future applicants. The NRC has issued three ESPs thus far, to System Energy

Resources, Inc., for the Grand Gulf site in Mississippi; to Exelon Generation Company, LLC, for the Clinton site in Illinois; and to Dominion Nuclear North Anna, LLC, for the North Anna site in Virginia. An ESP application from Southern Nuclear Operating Company for the Vogtle site in Georgia is currently under review.

The agency developed a Memorandum of Understanding (MOU) for consultation with the Department of Homeland Security (DHS) for new reactors under EPAct 2005 Section 657. This MOU was signed by the NRC and DHS in February 2007. Staff conducted monthly meetings with DHS to develop plans for implementation of the MOU and to ensure continuity of activities during this process.

In FY 2008, the NRC will continue to be a leader in developing programs to leverage the knowledge and resources within the international regulatory community in the licensing of new reactor designs. The NRC has played a key role in the Multinational Design Evaluation Program, an initiative through which several regulatory authorities share expertise and resources in reviewing new reactor designs, and are finding ways to harmonize codes, standards, and regulations for the review of future reactor designs.

In FY 2007, the NRC completed the first comprehensive review of Part 50 emergency preparedness regulations and guidance since the early 1980s, and has scheduled rulemaking, guidance, and generic communication modifications through 2010. The review incorporated extensive stakeholder involvement. Regulatory changes will codify prior orders, advisories, and information provided to licensees in response to the current threat environment.

The rising price of uranium has prompted considerable commercial interest in uranium production. In FY 2007, NRC staff was contacted by multiple companies stating their intent to submit up to 26 separate license applications for new in-situ leach (ISL) and conventional facilities, or for restarting or expanding existing facilities, in the FY 2007-2010 time frame. In FY 2007, in order to address the expected influx of new applications, the NRC developed and implemented an approach that focused on licensees developing high-quality applications, based on NRC staff facilitating early interactions and coordinating with interested parties. The staff requested that potential licensees provide letters of intent, indicating their plans to submit applications, to allow the staff to better plan and develop the infrastructure needed to support the application reviews. In FY 2007, the NRC received applications for expanding two facilities and restarting an existing facility. In early FY 2008, the NRC received one new application.

Licensing reviews for new uranium recovery facilities will be performed in accordance with applicable Standard Review Plans (SRPs) and other existing guidance, will include both a safety and an environmental review, and will be completed within a planned period of 24 months or less. In FY 2007, the staff developed an approach to prepare a Generic Environmental Impact Statement (GEIS) for ISL uranium recovery facilities. The GEIS will analyze the construction and operation of ISL facilities and discuss the potential environmental impact for reference resource areas that may be common to ISL facilities. Then, for each ISL facility application, a site-specific Environmental Assessment (EA) will be prepared that incorporates relevant conclusions from the GEIS and concentrates on potential environmental impacts that are unique to the proposed site. This approach should increase efficiency and minimize the potential for redundant and duplicative environmental analyses for proposed ISL facilities.

The agency expects to receive two license applications for new uranium enrichment plants in FY 2008. One proposed plant will utilize laser enrichment technology and the other will utilize gas centrifuge enrichment technology. NRC staff has completed two licensing reviews for uranium enrichment facilities in the past two years (Louisiana Energy Services and United States Enrichment Corporation, Inc.). In completing these reviews, the staff utilized a SRP. Following each review, the staff completed a lessons learned exercise identifying future improvements in the licensing process. These process improvements will be implemented in the upcoming reviews.

CHALLENGE 5

Implementation of information technology.

In FY 2007, the NRC began planning for the modernization of the Agencywide Documents Access and Management System (ADAMS). Staff established an ADAMS Governance Structure to provide strategic direction and NRC user input. Additionally, staff assessed the security controls of the current ADAMS to ensure they are consistent with both National Institute of Standards and Technology (NIST) guidance and the NRC's security policy. In addition, ADAMS was enhanced to facilitate the new reactor license review process. This enhancement added the functionality to view links between ADAMS documents and to copy groups of documents to a workstation. ADAMS will be further enhanced to allow internal users to access ADAMS documents using a Web-based interface from their workstations.

In FY 2008, the NRC plans to implement information technology and business process improvements to manage Requests for Additional Information (RAIs) associated with licensing reviews and to improve internal capability for collaboration among NRC staff. In addition, the NRC will streamline the process used by the public to search for documents stored in the ADAMS public library. Upon completion of this project, all publicly available ADAMS documents will be searchable from the public Web site, and the public will no longer need to access ADAMS to search for documents. Finally, the NRC will continue to evaluate the options for replacing the underlying technology supporting ADAMS. The result of this effort will be a defined path forward for improving ADAMS.

In FY 2007, the first phase of the New Reactor Application Document Intake and Review Pilot was deployed to provide the capability for applicants to create and provide electronic Combined Operating License (COL) submittals to the NRC. Westinghouse successfully submitted a new reactor design control document (DCD) to the NRC in May 2007. The Westinghouse DCD consisted of 270 individual files with navigational links and was profiled into ADAMS in 2 hours. Manual processing of this DCD would have taken two days (160 work-hours) and would not have supported the navigational links. NRC staff consolidated guidance on how to submit documents electronically to the agency and issued it for public comment. The consolidated guidance includes a new chapter addressing COL submittals.

In FY 2007, the NRC developed and implemented information technology and business process improvements to automate the capture of e-mail comments into ADAMS as Official Agency Records on NRC proceedings such as the North Anna Early Site Permit. This information technology solution has also been configured to automate capture of e-mail relevant to hearing files in preparation for reactor license renewal hearings. In combination with business process improvements, this automated e-mail capture solution is allowing staff to effectively meet the challenges presented by high volumes of comment and discovery documents. During the initial comment period for the Vogtle Early Site Permit, savings of over 500 work-hours were identified by automating the processing of comments into ADAMS. The NRC achieved a full return on this investment in the first three months of implementation.

The NRC has streamlined the business process and improved the information technology supporting general adjudicatory hearings and the E-Filing rule. This rule, which will go into effect in FY 2008, will codify electronic filing and conduct of agency adjudicatory licensing hearings, including the hearings for reactor license renewals, materials licensing, and new reactor licensing.

In February 2007, staff completed a redesign of the NRC's public Web site. This redesign incorporated a new NRC banner and site-wide menu system, streamlined site content categories, and added a Google search function at the top of each page.

CHALLENGE 6

Administration of all aspects of financial management.

In FY 2007, for the sixth consecutive year, the NRC received the Certificate of Excellence in Accountability Reporting (CEAR Award) for the agency's Performance and Accountability Report. The CEAR Program, sponsored by the Association of Government Accountants, was established in conjunction with the Chief Financial Officers Council and the Office of Management and Budget. Its goal is to improve financial and program accountability by streamlining reporting and improving the effectiveness of such reports.

In FY 2007, the NRC received an unqualified audit opinion on its FY 2006 financial statements. The agency's independent auditors continued to characterize the NRC's legacy fee billing system as a material weakness and as a Federal Financial Management Improvement Act substantial non-compliance. The age of the system, reliance on manual processes, and lack of comprehensive quality assurance procedures are the underlying cause of the material weakness. The NRC is addressing the continuing material weakness by assessing all processes and system interfaces associated with the fee billing process and system to ensure controls are adequate. The agency has implemented a number of new and improved controls to include a validation tool which analyzes and reconciles the completeness and accuracy of billing for reactors and material inspections. As result, the agency has decreased the risk of potential billing errors and further enhanced the control environment.

In FY 2007, the NRC developed a number of internal control tools to mitigate the effects of these deficiencies. Improvements included further automating processes, conducting self and OMB Circular A-123 Appendix A assessments, standardizing billing certification procedures, strengthening quality control reviews, and conducting program reviews over certification processes.

The financial systems replacement project, as currently planned, involves the replacement and integration of the NRC's core accounting system, license fee billing system, cost accounting system, allotment/allowance financial plan system, and capitalized property system. During FY 2007, the agency developed the business case for its new financial system, for which final approval is expected by the end of the fiscal year. This included researching shared service providers and commercial off the shelf software for potential use. The new financial system has a target implementation date of October 1, 2009.

NRC conducted a business process improvement study focused on time and labor and fee billing processes. The study has made a number of recommendations for improvement including the need to corporately manage the reporting codes and to reduce the number of reporting codes to improve internal controls. To date, as a result, the agency has developed interim guidance for managing reporting codes and will reduce the number of codes by over 30 percent by year end.

In FY 2008, the agency is moving forward with the upgrade of the Time and Labor portion of its human resource management system. The upgraded system is due to be completed, including electronic signature capabilities, in FY 2009.

The NRC also procured a vendor to integrate a new Web-based budget system and tested it as a pilot during the FY 2009 PBPM process. The required system certification and accreditation is in its final review, and the approval and authorization to operate the new system is expected in FY 2008.

CHALLENGE 7

Communication with external stakeholders throughout NRC regulatory activities.

During FY 2007, the NRC continued to reach out and to work collaboratively with its many Native American Tribal stakeholders who have expressed concern and interest regarding NRC-related licensing activities. Throughout the year, representatives from the National Congress of American Indians (NCAI) met with NRC staff, as well as Chairman Klein, in order to discuss increased outreach and inclusiveness of Native American tribes on nuclear issues. Chairman Klein reinforced NRC's commitment to support and encourage Native American students to pursue advanced education in engineering and the sciences. In support of that commitment, NRC sponsored four Native American student interns through American University's Washington Internships for Native Students program. NRC staff also participated in several conferences in order to strengthen relations with Tribal stakeholders, such as the U.S. Department of Energy's Tribal Workshop held in Denver, Colorado, and the NCAI Mid-Year Conference in Anchorage, Alaska. Discussions focused on a variety of issues including economic development, natural resources, education, and the preservation of Native cultures. NRC conducted one of a series of scoping meetings on the Generic Environmental Impact Statement for Uranium Recovery Licensing in Gallup, New Mexico. This meeting included a number of Navajo participants, and NRC's presentations were translated into the Navajo language by a Navajo translator.

In FY 2007, the public was able to keep abreast of the NRC's reactor license renewal regulatory activities through a variety of open meetings, including advisory committee meetings and staff meetings open to the public. The NRC holds open meetings in the vicinity of plants undergoing license renewal reviews. These meetings are held to describe the license renewal process, solicit environmental scoping

issues, and to obtain comments on the environmental impact statement developed by the NRC. In addition, the license renewal section of the NRC public Web site describes the process, regulations, guidance, opportunities for public involvement, and status of current activities associated with renewal of licenses for commercial operating power reactors.

The NRC also continued an active stakeholder outreach program on spent fuel storage and transport. In addition to outreach on specific licensing requests and interaction with the industry, the NRC participated in meetings of the Northeast Governors Task Force, the Midwest States Task Force, the Western Governors Association, the Western Interstate Energy Board, and the Southern States Energy Board, and participated in multiple local and national meetings and conferences to discuss the NRC's safety regulations for the transportation and storage of spent nuclear fuel.

In June 2007, the NRC hosted a two-day seminar, the Fuel Cycle Information Exchange. This seminar provided an opportunity for licensees, NRC staff, and other stakeholders to exchange information and discuss issues of interest pertaining to the regulation of NRC-licensed fuel cycle facilities. The NRC is also making publicly available documents that were previously withheld regarding certain fuel cycle licensees, so that the public can remain aware of NRC activities at those sites.

The NRC also increased its outreach activities to external stakeholders in the areas of emergency preparedness and incident response. Stakeholder involvement in the emergency preparedness regulatory review and increased involvement with Federal, State, local, and Tribal incident response organizations through exercise participation are examples where the NRC broadened its outreach scope to enhance programs.

In FY 2007, interest in the uranium recovery area was high due to increasing uranium prices. The NRC held 17 public meetings regarding the uranium recovery area. In the decommissioning area, the NRC held a public roundtable meeting to discuss potential rulemaking to reduce the likelihood of funding shortfalls for decommissioning under the license termination rule. Approximately 70 stakeholders and interested members of the public participated in this discussion. NRC staff also conducted an informal public meeting to discuss the decommissioning of the Shieldalloy Metalurgical Corporation site in Newfield, New Jersey.

The NRC continued to hold monthly conference calls with members of the Organization of Agreement States and the Conference of Radiation Control Program Directors to inform the States and glean the States' views on issues affecting them. Other calls were held with the respective executive boards of these organizations on urgent special topics of interest to the States.

The staff continued to emphasize stakeholder involvement and open communication regarding the Reactor Oversight Process (ROP). The staff used a variety of communication methods to ensure that all stakeholders had access to ROP information and were given an opportunity to participate in the process and provide feedback. The staff conducted monthly public working-level meetings with the Nuclear Energy Institute (NEI), the industry, and interested stakeholders to discuss ongoing refinements to the ROP. The staff also conducted public meetings in the vicinity of each operating reactor to discuss the results of the NRC's annual assessment of the licensee's performance. These meetings provided interested stakeholders with an opportunity to engage the NRC on plant performance and the role of the agency in ensuring safe plant operations. The staff sponsored a session at the Regulatory Information Conference (RIC) in March 2007 on the ROP, with a focus on implementation of safety culture enhancements, and discussed additional ROP topics during the regional breakout sessions. The staff also issued its annual external survey through the *Federal Register* in October 2006 to evaluate ROP effectiveness and gather stakeholder insights. Finally, the staff maintained and enhanced the NRC's Web pages to communicate current ROP-related information

and results. These outreach efforts have resulted in valuable feedback and ROP improvements.

Subsequent to the publication of the Power Reactor Security Requirements, at the beginning of FY 2007, several public meetings were held in various locations to provide an opportunity for the public to provide comments on the proposed rule and for licensees, other stakeholders, and NRC staff to exchange information on interpretation and implementation of specific areas of proposed regulations. When the public comment period closed, the agency had received over 600 pages of comments. Staff plans to conduct public meetings following the publication of draft guidance supporting this rule to facilitate understanding and allow interested stakeholders an opportunity to ask questions.

CHALLENGE 8

Managing human capital.

The NRC was ranked as the Best Place to Work in the Federal Government in the Office of Personnel Management's (OPM's) 2006 Federal Human Capital Survey. These rankings have created a baseline for measuring employee commitment and engagement and are being used to identify and concentrate on areas for improvement. As part of this continuous improvement effort and in light of NRC's current shifting age demographic, a series of staff focus groups will be conducted to go "behind the numbers" in the OPM survey to help understand the agency's strengths and areas for improvement in terms of attracting and retaining a highly skilled, engaged workforce. NRC staff members are conducting these focus groups, compiling the information gathered, and will present a proposed action plan to senior managers in FY 2008.

The NRC exceeded its FY 2006 hiring goal and is on track to exceed its FY 2007 goal to reach a net gain of approximately 200 staff. To achieve this goal, the

NRC maintains a vigorous recruitment program that includes participation in approximately 80 recruitment events each year at colleges, universities, and professional conferences as well as career invitational events. In FY 2007, the NRC began efforts to improve the recruitment and staffing program through: developing new recruitment displays and three videos of selected NRC employees to show at recruitment events; hiring additional staff to perform critical human resources work; and upgrading the agency's Web-based job application tool to gain more functionality.

The NRC continued its use of recruitment and relocation incentives, credit for non-Federal service toward annual leave earnings, salary exceptions, student loan repayment program and the provision of student lodging and transportation expenses to attract highly qualified candidates. The NRC implemented an expedited process to review a broader range of incentives in order to meet temporary emergency hiring needs and/or recruit and retain individuals in positions for which exceptional recruitment or retention difficulty exists. The NRC also continues to provide a wide range of flexible work options and employee-friendly programs and policies to attract and retain employees.

In FY 2007, the NRC used the Lean Six Sigma methodology to evaluate and develop recommendations for streamlining the hiring process (measured from the closing date of a vacancy announcement to the date an offer is extended). Efforts are underway to implement the recommendations made by the Lean Six Sigma workgroup and to develop a plan to assess the NRC's progress towards reducing the hiring time frame to meet a target of 45 days.

NRC offices and regions continued to play an important role in supporting the NRC's recruitment efforts to achieve a high quality, diverse workforce. For example, Region III implemented a successful recruitment strategy that included: recruiting at local colleges and universities; strengthening and developing relationships with targeted groups; and

expanding outreach to selected secondary and elementary schools in the Chicago area to encourage children to pursue careers in engineering and science through age-appropriate presentations in the classroom. The Office of Nuclear Material Safety and Safeguards (NMSS) reached out to the Nuclear Engineering Department Heads Organization and briefed the group at their June 2007 meeting in Boston. NMSS representatives took this opportunity to present NRC recruiting strategies and goals for FY 2008. Further, NMSS collaborated with the Office of Human Resources to develop a plan to obtain and sustain various technical expertise during the future review of the Yucca Mountain license application.

The Office of Human Resources partnered with the Office of New Reactors in FY 2007 to develop and deliver technical, professional development, and computer skills courses (e.g., advanced courses on the AP1000 and Advanced Boiling Water Reactor designs) for personnel involved in the review of new reactor licensing applications. In addition, the NRC is increasing space available in the Professional Development Center to provide two additional classrooms and technical training aid display areas to support increased technical and professional training for the new reactor program.

In FY 2007, the NRC continued to develop a Knowledge Management (KM) program to integrate new and existing approaches for generating, capturing, and transferring knowledge relevant to the NRC's mission. Currently, the agency's knowledge-sharing practices include formal and informal mentoring, storytelling, early replacement hiring, and rehiring annuitants expressly for the purposes of retaining knowledge and/or recovering lost knowledge. In addition, the agency continued to develop Communities of Practice ("CoPs"): groups of individuals who regularly interact to share knowledge regarding a particular topic, method, or work role. The agency is in the process of implementing virtual CoPs to provide communities with an online environment supported through commercial CoP software tools to facilitate knowledge sharing.

MANAGEMENT DECISIONS AND FINAL ACTIONS ON OIG AUDIT RECOMMENDATIONS

The agency has established and continues to maintain an excellent record in resolving and implementing audit recommendations presented in OIG reports. Section 5(b) of the Inspector General Act of 1978, as amended, requires agencies to report on final actions taken on OIG audit recommendations. The following table gives the dollar value of disallowed costs determined through contract audits conducted by the Defense Contract Audit agency and NRC's Office of the Inspector General. Because of the sensitivity of contractual negotiations, details of these contract audits are not furnished as part of this report. As of September 30, 2007, there were no outstanding audits recommending that funds be put to better use.

MANAGEMENT REPORT ON OFFICE OF THE INSPECTOR GENERAL AUDITS WITH DISALLOWED COSTS FOR THE PERIOD OCTOBER 1, 2006–SEPTEMBER 30, 2007

Category	Number of Audit Reports	Questioned Costs	Unsupported Costs
1. Audit reports with management decisions on which final action had not been taken at the beginning of this reporting period.	0	$0	$0
2. Audit reports on which management decisions were made during this period.	1	$193,585	$0
3. Audit reports on which final action was taken during this report period.			
(i) Disallowed costs that were recovered by management through collection, offset, property in lieu of cash, or otherwise.	0	$0	$0
(ii) Disallowed costs that were written off by management.	0	$0	$0
4. Reports for which no final action had been taken by the end of the reporting period.	0	$193,585	$0

MANAGEMENT DECISIONS NOT IMPLEMENTED WITHIN ONE YEAR

Management decisions were made before October 1, 2006, for the OIG audit reports listed in the following tables. As of September 30, 2007, NRC did not take final action on some issues. Completion of the activities listed as "Actions Pending" will complete agency action on the listed OIG audit and evaluation recommendations.

GOVERNMENT PERFORMANCE AND RESULTS ACT: REVIEW OF THE FY 1999 PERFORMANCE REPORT (OIG-01-A-03) FEBRUARY 23, 2001

This audit was conducted at the request of the Chairman of the Senate Committee on Governmental Affairs to determine if NRC's FY 1999 performance data was valid and reliable and if the FY 2000 performance data would be more valid and reliable. The audit found that while NRC was improving and strengthening its performance reporting process, management control procedures required to produce valid and reliable data needed to be put in place as interim policy guidance and then institutionalized in an NRC management directive.

Open Recommendations	Actions Pending
1. Develop an NRC management directive (MD) to provide the management controls needed to ensure that NRC produces credible Government Performance and Results Act (GPRA) documents. 3. Include guidance on reporting unmet goals in both the management directive and the interim policy guidance on implementing GPRA initiatives.	Interim guidance for performance management and reporting performance information was issued in July 2001. In July 2002, a new MD and Handbook 4.8, Performance Measurements, was issued for intra-agency review and comment. It was subsequently decided that performance measurement should be addressed in the broader context of budget and performance integration. Therefore, new MD 4.8 is being incorporated into a revision of MD and Handbook 4.7, which will be entitled Planning, Budgeting, and Performance Management. Revised MD 4.7 will clarify the roles and responsibilities in setting the agency's strategic direction, determining planned activities and resources, measuring and monitoring performance, and assessing performance. The revised management directive and handbook is expected to be issued in February 2008.

REVIEW OF THE AGENCYWIDE DOCUMENTS ACCESS AND MANAGEMENT SYSTEM (OIG-02-A-12) JUNE 12, 2002

This audit was conducted to determine how effectively NRC carried out the Chairman's request for an assessment of the effectiveness and efficiency of the Agencywide Documents Access and Management System (ADAMS), the electronic system that maintains official NRC records, and to assess what additional NRC actions are required to make ADAMS successful. The audit found that NRC needed to improve ADAMS management controls.

Open Recommendation	Action Pending
1. Finalize and issue Management Directive (MD) 2.5, Application Systems Life-Cycle Management and Handbook 2.5, System Development and Life-Cycle Management Methodology.	A new MD 2.8, Project Management Methodology– superceding MD 2.1, Information Technology Architecture, MD 2.2, Capital Planning and Investment Control, and previously issued draft MD 2.5–was issued in June 2007. OIG's review of NRC actions taken and closure was pending at the time of the FY 2007 Performance and Accountability Report's preparation.

REVIEW OF NRC'S HANDLING AND MARKING OF SENSITIVE UNCLASSIFIED INFORMATION (OIG-03-A-01) OCTOBER 25, 2002

This audit was conducted to assess NRC's program for the handling, marking, and protection of Official Use Only (OUO) information, a category of sensitive unclassified information. The audit found that NRC's program and guidance for the handling and marking of sensitive unclassified information may not adequately protect OUO information from inadvertent public disclosure and that training on handling and protecting sensitive unclassified information is not provided to all NRC employees and contractors on a regular basis.

Open Recommendations

1. Update the guidance for OUO documents to require clear identification of sensitive unclassified information to prevent its inadvertent disclosure.

2. Mandate consistent use of defined markings on documents containing OUO information and clarify the markings that should be used on sensitive unclassified information.

Actions Pending

Agency corrective actions require issuance of a revised management directive (MD) covering sensitive unclassified, non-safeguards information (SUNSI) and a new MD covering safeguards information (SGI). It is expected that the new MD on SGI will be issued by December 2007. With respect to SUNSI, the staff is developing a proposed policy, which is scheduled to be provided to the Commission for review and approval by the end of June 2008. Following receipt of the Commission's guidance on the proposed policy, the staff will develop the revised MD on SUNSI, which is expected to be issued by the end of December 2008.

USE OF ELECTRONIC MAIL AT NRC (OIG-03-A-11) MARCH 21, 2003

This audit was conducted to determine whether NRC has an adequate process for ensuring that appropriate items of electronic mail (e-mail) correspondence become official agency records, adequate policies and procedures covering the use of its e-mail system, and employee and contractor use of the e-mail system is consistent with agency policy. The audit found that adequate controls for ensuring that appropriate e-mail records become official agency records have not been implemented, and while NRC employees generally use the e-mail system for official business or limited personal use in accordance with agency policy, contractors do not follow the more stringent e-mail usage policy applicable to them.

Open Recommendation

1. Revise "Management Directive and Handbook 3.53, NRC Records Management Program," to include current information about capturing e-mail records in the Agencywide Documents Access and Management System (ADAMS).

Action Pending

The revised management directive and handbook was issued in March 2007. OIG's review of NRC actions taken and closure was pending at the time of the FY 2007 Performance and Accountability Report's preparation.

AUDIT OF NRC'S REGULATORY OVERSIGHT OF SPECIAL NUCLEAR MATERIALS
(OIG-03-A-15) JUNE 3, 2003

This audit was conducted to determine whether NRC adequately ensures its licensees control and account for special nuclear material (SNM). The audit found that NRC's current levels of oversight of licensees' material control and accounting (MC&A) activities do not provide adequate assurance that all licensees properly control and account for SNM in that NRC performs only limited inspections of licensees' MC&A activities and cannot assure the reliability of data in the Nuclear Materials Management and Safeguards System, which is a computer database managed by the U.S. Department of Energy and jointly used with NRC as the national system for tracking certain private- and Government-owned nuclear materials.

Open Recommendations

1. Conduct periodic inspections to verify that material licensees comply with MC&A requirements, including but not limited to visual inspections of licensees' SNM inventories and validation of report information.

3. Document the basis of the approach used to risk-inform NRC's oversight of MC&A activities for all types of materials licensees.

4. Revise NRC regulations to require licensees authorized to possess SNM, and not currently required to do so, to conduct annual inventories and submit an annual Material Status Report or Physical Inventory Summary Report to NRC.

Actions Pending

NRC expects to issue a proposed rule by June 2009, with issuance of the final rule by December 2010, to make enhancements to MC&A regulations, inspections, and licensing process. The work on this rulemaking will include documentation of the technical basis for risk-informing the MC&A program and how it will be applied to the program. By July 2011, NRC expects to have completed the application of risk-informing the MC&A program with respect to determining inspection resources and frequencies for all types of materials licensees MC&A inspections for SNM.

A proposed rule was published for public comment in February 2007. NRC expects to issue a final rule in the spring of 2008, to require all licensees possessing one gram or more of SNM to submit a completed Material Status Report and Physical Inventory Listing to NRC annually.

REVIEW OF NRC'S DRUG-FREE WORKPLACE PLAN (OIG-04-A-15) MAY 24, 2004

The audit of NRC's Drug Testing Program (discussed further in the table on OIG-05-A-05) found that the NRC's Drug-Free Workplace Plan was not in compliance with Federal guidance that requires the plan to receive U.S. Department of Health and Human Services' (HHS's) approval and that it was missing a required clause.

Open Recommendations

1. Revise the NRC Drug-Free Workplace Plan to include the deferral-of-testing clause from the HHS's Model Plan for a Comprehensive Drug-Free Workplace Program.

2. Include in the NRC Drug-Free Workplace Plan instruction that revisions must receive approval from the HHS prior to implementation.

3. Obtain HHS's approval of the 2004 NRC Drug-Free Workplace Plan prior to implementation.

Actions Pending

The plan was revised to include the deferral-of-testing clause and an instruction that plan revisions must receive approval from HHS prior to implementation. HHS approved the NRC's plan on August 24, 2007. NRC considers action in response to these recommendations to be complete, although closure requires OIG's review of HHS's approval of the final updated plan, which is expected in early FY 2008.

AUDIT OF NRC'S INCIDENT RESPONSE PROGRAM (OIG-04-A-20)   SEPTEMBER 23, 2004

This audit was conducted to determine whether NRC's incident response program is performed in a timely and effective manner, provides adequate support to licensees, and maintains readiness and qualifications of staff. The audit found that while NRC has improved its program since the Three Mile Island 2 accident on March 29, 1979, more needed to be done to ensure that the program is performed consistently, is more fully understood by licensees, and maintains a well-defined process for demonstrating staff are qualified and ready to respond.

Open Recommendations	Actions Pending
3. Update NUREG-0845, "Agency Protocols for the NRC Incident Response Plan," or incorporate relevant portions into other agency procedures.	NUREG-0728, "NRC Incident Response Plan," Revision 4, which was issued for interim use effective April 14, 2005, supersedes NUREG-0845. The relevant portions of NUREG-0845 have been incorporated into NUREG-0728 and its implementing procedures. All revised implementing procedures are expected to be completed by September 1, 2008.
4. Periodically review regional incident response programs to ensure NRC's incident response program is carried out consistently across the agency.	Implementation of the Incident Response Self-Assessment Program began with the development of a draft self-assessment plan that was tested in NRC Region II during the week of October 15, 2007, in concert with the McGuire full-participation exercise. Plans for instituting self-assessments in all of the NRC Regions are expected to be completed by December 2008.
8. Periodically conduct incident response exercises involving multiple sites.	In order to test the key elements of Incident Response Manual Chapter (IRMC) 0920, "Incident Response-Multiple Incidents," NRC conducted a multiple-event tabletop exercise in March 2006 that included participation by all of the NRC regional offices. On March 21, 2007, during the Dresden Exercise, the NRC successfully demonstrated its ability to conduct a multiple-incident exercise. After further review, the NRC has determined that IRMC 0410, "NRC Drill and Exercise Standards," is the appropriate document to capture the requirement. IRMC 0410 will be revised by January 31, 2008, to include a requirement for multi-incident exercises once per year.
11. Revise the NRC Incident Response Plan to better define the incident response to emergencies involving regulated fuel cycle facilities and nuclear materials.	After review of the "NRC Incident Response Plan" (NUREG-0728), the NRC staff determined that the IRMCs for fuel cycle facilities and subsequent response procedures are the appropriate location for incident response guidance on fuel cycle facilities. The IRMCs for fuel cycle facilities and associated response procedures are currently under construction and are expected to be issued by March 31, 2008.
13. Update response technical manual (RTM) supplements for gaseous diffusion plants.	The NRC staff has begun an effort to update the RTM supplements for the gaseous diffusion plants (GDPs). The effort is focused on evaluating pertinent information, such as locations and quantities of uranium hexaflouide (UF6), which would affect the NRC's event assessments should an accident occur. After site visits to the GDPs in May 2007, the staff began to evaluate the information in the current RTM supplements and UF6 inventories at the plants in order to determine whether related information in the RTM supplements would be effective for event assessment activities. The results of the evaluation will be used to identify what revisions to the RTM supplements are warranted. Any revisions are expected to by completed by December 2007.

SYSTEM EVALUATION OF THE AGENCYWIDE DOCUMENTS ACCESS AND MANAGEMENT SYSTEM (OIG-04-A-21) OCTOBER 21, 2004

This evaluation was conducted as part of the OIG's review of NRC's implementation of the Federal Information Security Management Act (FISMA) for FY 2004, with the objectives of reviewing and evaluating the management, operational, and technical controls for NRC's Agencywide Documents Access and Management System (ADAMS). The review found that ADAMS security documentation was not always consistent with National Institute of Standards and Technology (NIST) guidelines, security protection requirements were not consistent within the security documentation, and findings and recommendations resulting from testing were not consistently tracked.

Open Recommendations	Actions Pending
1. Update the ADAMS Risk Assessment Report to be consistent with NIST Special Publication 800-30, "Risk Management Guide."	The ADAMS Risk Assessment Report was updated as part of the ADAMS security certification and accreditation to be consistent with the applicable NIST and NRC guidance. The final ADAMS Risk Assessment Report was approved by the NRC's Senior Information Technology Security Officer (SITSO) on January 23, 2007. NRC considers action in response to this recommendation to be complete, although closure requires OIG's verification that the ADAMS Risk Assessment Report is consistent with NIST Special Publication 800-30, and is expected in early FY 2008.
2. Update the ADAMS Security Plan to describe all controls currently in place. In-place controls are those marked at least at Level 3 in the self-assessment and that were documented as passed in the last Security Test and Evaluation Report or in any test and evaluation on controls added since publication of that report.	Update of the ADAMS Security Plan was dependent on a completed ADAMS Risk Assessment Report. The final ADAMS Risk Assessment Report was approved by the NRC's SITSO on January 23, 2007. The ADAMS Security Plan was updated as part of the ADAMS security certification and accreditation, to describe all controls in place. The ADAMS Security Plan was completed on August 8, 2007. NRC considers action in response to this recommendation to be complete, although closure requires OIG's verification that the final updated plan describes all controls in place, and is expected in early FY 2008.
4. Update the ADAMS Business Continuity Plan to include the following changes: • Describe the methods used to notify recovery personnel during business and non-business hours. • Incorporate all teams' roles and responsibilities and relevant points of contact information for team leaders, alternate team leaders, and team members for all scenarios. • Include procedures for restoring system operations with a focus on how to clean the alternate site of any equipment or other materials belonging to the organization.	The ADAMS Business Continuity Plan (BCP) is being updated as part of the ADAMS certification and accreditation, and will include the recommended changes. The BCP is dependent on a completed and approved security plan. The BCP is expected to be updated by November 1, 2007.

continued

SYSTEM EVALUATION OF THE AGENCYWIDE DOCUMENTS ACCESS AND MANAGEMENT SYSTEM (OIG-04-A-21), CONTINUED

Open Recommendations	Actions Pending
5. Update the ADAMS Security Plan and/or ADAMS self-assessment to consistently define the protection requirements (confidentiality, integrity, availability).	Update of the ADAMS Security Plan was dependent on a completed ADAMS Risk Assessment Report. The final ADAMS Risk Assessment Report was approved by the NRC's SITSO on January 23, 2007. The ADAMS Security Plan was updated as part of the ADAMS security certification and accreditation to define all the controls currently in place, including the protection requirements. The ADAMS Security Plan was completed on August 8, 2007. NRC considers action in response to this recommendation to be complete, although closure requires OIG's verification that the ADAMS Security Plan and/or ADAMS self-assessment consistently defines the protection requirements, and is expected in early FY 2008.
6. Track all action items resulting from testing of the ADAMS security controls and contingency plan in either the agency's internal tracking system or the agency's plan of action and milestones (POA&M).	All items resulting from testing of the ADAMS security controls and contingency plan will be placed in either Rationale Clear Case, NRC's internal tracking system, or in the NRC's FISMA plan of action and milestones (POA&M) submitted to OMB. Rational Clear Case will be updated to track all action items as results become available. The test results are expected to be documented by November 1, 2007.

INDEPENDENT EVALUATION OF NRC'S IMPLEMENTATION OF THE FEDERAL INFORMATION SECURITY MANAGEMENT ACT FOR FY 2004 (OIG-04-A-22) SEPTEMBER 30, 2004

This was an independent evaluation of NRC's implementation of the Federal Information Security Management Act for FY 2004. The review found that while NRC had made improvements to its automated information security program, additional improvements were needed.

Open Recommendations	Actions Pending
Five of the original 16 recommendations remain open.	Due to the sensitive nature of the OIG's review and recommendations in this area, specific details are not furnished as part of this report. As of September 30, 2007, completion of agency actions on this OIG audit report requires re-certification and re-accreditation of some systems and updating of a business continuity plan. These action are being completed in accordance with a prioritization of information technology security activities, which is based on a mission perspective and security risk. Consequently, most of these activities are expected to be completed in the first half of FY 2008, but completion of the re-certification and re-accreditation work will be delayed until early FY 2009. These agency actions will be carried over to and tracked to completion via NRC's FY 2008 Plan of Action and Milestones required by the Federal Information Security Management Act.

SYSTEM EVALUATION OF THE GENERAL LICENSE TRACKING SYSTEM
(OIG-04-A-24) OCTOBER 21, 2004

This evaluation was conducted as part of the OIG's review of NRC's implementation of the Federal Information Security Management Act for FY 2004, with the objectives of reviewing and evaluating the management, operational, and technical controls for the General License Tracking System (GLTS), the primary function of which is to facilitate the tracking and accountability of NRC general licensees and generally licensed devices. The review found that the GLTS's security documentation did not always follow required guidelines, security protection requirements were not consistent within the security documentation, and NRC was not tracking all action items resulting from testing the system's security controls.

Open Recommendations	Actions Pending
1. Update the GLTS Security Plan to describe all controls currently in place. In-place controls are those marked at least at Level 3 in the self-assessment and that were documented as passed in the last Security Test and Evaluation Report, or in any test and evaluation on controls added since publication of that report.	A task order was issued for the completion of the certification and accreditation process and preparation of deliverables in compliance with National Institute of Standards and Technology (NIST) and Federal Information Security Management Act (FISMA) guidance for the GLTS. The initial kick-off meeting for this work is expected to be held by December 31, 2007. One product required of the contractor at the kick-off is a draft project plan and schedule. Completion of the kick-off, project plan, and schedule will provide the target dates for completion of each security artifact, including the GLTS System Security Plan (SSP). The updated SSP resulting from this effort will take into account the requirements for full description of all in-place controls. However, the period of performance for the task order is through September 25, 2008, by which time it is expected that the updated SSP will be completed.
3. Update the GLTS Business Continuity Plan.	The task order discussed above for Recommendation 1 was issued to accomplish deliverables in compliance with NIST and FISMA guidance. The completion date for the updated GLTS Contingency (Business Continuity) Plan will not be known until the kick-off meeting, which is expected to be held by December 31, 2007. The Contingency Plan can be completed up to 180 days after receiving the Authority to Operate (ATO) the system. Completion of the kick-off meeting and documentation leading to a successful ATO will determine the target date for completion of the update of the GLTS Contingency Plan. However, the period of performance for the task order is through September 25, 2008, by which time it is expected that the updated Contingency Plan will be completed.
4. Update the GLTS Security Plan and/or GLTS self-assessment to consistently define the protection requirements (confidentiality, integrity, availability).	During development of the updated GLTS SSP, the contractor will be alerted to the need to ensure consistency in defining the protection requirements. As discussed above for Recommendation 1, completion of the kick-off, project plan, and schedule will provide the target dates for completion of each security artifact, including the GLTS System Security Plan (SSP). However, the period of performance for the task order is through September 25, 2008, by which time it is expected that the updated SSP will be completed.

AUDIT OF NRC'S DRUG TESTING PROGRAM (OIG-05-A-05)   DECEMBER 30, 2004

This audit was conducted to assess the NRC's implementation of its drug testing program, and identified that improvements were needed in the program's random testing process and management oversight.

Open Recommendations	Actions Pending
4. Revise the categories of testing-designated positions to include computer system administrators and individuals engaged in law enforcement activities who are authorized to carry weapons.	On September 29, 2006, the Commission decided to revise the drug testing pool to include all NRC employees. Appropriate changes were incorporated in the NRC Drug-Free Workplace Plan to reflect this decision and the plan was submitted to the U.S. Department of Health and Human Services (HHS) for review and approval. HHS approved the NRC's plan on August 24, 2007. As a result of this approval, the staff has prepared a revised implementation plan (with dates for various actions to be completed) which calls for implementation of the revised NRC Drug-Free Workplace Plan 250 days after HHS's approval of the plan, i.e., by August 25, 2008. Closure of this recommendation requires OIG verification that NRC has implemented the revised NRC Drug-Free Workplace Plan with this provision.
5. Re-evaluate categories of testing-designated positions and continue to do so biennially.	The NRC Drug-Free Workplace Plan was approved by HHS on August 24, 2007, and provides for the testing-designated position criteria to be reviewed and revised as appropriate on a biennial basis. Closure of this recommendation requires OIG verification that NRC has implemented the revised NRC Drug-Free Workplace Plan with this provision.
12. Update the Management Directive System to include the drug testing policy and procedures that employees are expected to follow.	With HHS's approval of NRC's plan on August 24, 2007, a new management directive (MD) that will describe the NRC's drug testing policy and provide an overview of the procedures that employees are to follow is expected to be issued by August 2008. Closure of this recommendation requires OIG evaluation of the revised MD containing the drug testing policy and procedures that employees are expected to follow.

SYSTEM EVALUATION OF THE INTEGRATED PERSONNEL SECURITY SYSTEM
(OIG-05-A-08) JANUARY 26, 2005

This evaluation was conducted as part of the OIG's review of NRC's implementation of the Federal Information Security Management Act for FY 2004, with the objectives of reviewing and evaluating the management, operational, and technical controls for the Integrated Personnel Security System (IPSS), which replaced NRC employee security information contained in paper files and in a less-capable automated data system. The review found that the IPSS's security test and evaluation were not comprehensive and independent, security documentation was not always consistent with National Institute of Standards and Technology (NIST) guidelines, and security protection requirements were not consistent within the security documentation.

Open Recommendations	Actions Pending
1. Re-certify and re-accredit IPSS based on an independent, comprehensive, and fully documented assessment of all management, operational, and technical controls.	Completion dates have been established in order to integrate the certification and accreditation of IPSS with the implementation of Homeland Security Presidential Directive 12 and to allow time for resolution of operational issues. Therefore, certification and accreditation of IPSS is expected to be completed by March 31, 2008.
2. Update the IPSS Risk Assessment Report to include listed changes.	The IPSS Risk Assessment Report is expected to be updated to include the specified items by December 31, 2007.
3. Update the IPSS System Security Plan to include listed changes.	The IPSS Security Plan is expected to be updated to include the specified items by December 31, 2007.
4. Update the IPSS System Security Plan to include a section on planning for security in the life cycle and a section on incident response capability.	The IPSS Security Plan is scheduled to be updated by December 31, 2007 and will include sections on planning for security in the life cycle and incident response capability.
5. Update the IPSS System Security Plan to describe all controls currently in place. In-place controls are those marked at least at Level 3 in the self-assessment and that were documented as passed in the last Security Test and Evaluation Report, or in any test and evaluation on controls added since publication of that report.	The IPSS Security Plan is expected to be updated by December 31, 2007 and will describe all controls currently in place.
7. Update the IPSS Contingency Plan to include listed changes.	The IPSS Contingency Plan is expected to be updated by December 31, 2007 to include the specified items.
8. Update the IPSS System Security Plan and/or IPSS self-assessment to consistently define the protection requirements (confidentiality, integrity, availability).	The security plan and IPSS self-assessment are expected to be updated by December 31, 2007 to consistently define protection requirements.

AUDIT OF NRC'S BUDGET FORMULATION PROCESS (OIG-05-A-09) FEBRUARY 9, 2005

This audit was conducted to determine whether the budget formulation portion of the NRC's Planning, Budgeting, and Performance Management process is effectively used to develop and collect data to align resources with strategic goals and efficiently and effectively coordinated with program and support offices. The audit identified that NRC effectively develops and collects data to align resources with strategic goals, prepares the budget in alignment with the Strategic Plan, and successfully conducts Office of Management and Budget-required Program Assessment Rating Tool evaluations, but needed additional internal coordination and communication efforts.

Open Recommendations	Actions Pending
1. Clarify the roles and responsibilities of the Chief Financial Officer and the Executive Director for Operations in the budget formulation process.	In July 2007, the NRC staff provided a Program Review Committee Charter to the Commission for review and approval. A revision of Management Directive 4.7, Planning, Budgeting, and Performance Management, will clarify roles and responsibilities and document the budget formulation process, including decision-making, and will provide for a logical, comprehensive sequencing of events for obtaining early Commission direction and approval. The revised management directive and handbook is expected to be issued in February 2008.
2. Document the decision-making process and roles and responsibilities of the Program Review Committee.	
3. Document the budget formulation process to ensure a logical, comprehensive sequencing of events that provides for obtaining early Commission direction and approval.	

AUDIT OF NRC'S TELECOMMUNICATIONS PROGRAM (OIG-05-A-13) JUNE 7, 2005

This audit was conducted to evaluate controls over the use of NRC telecommunications services and the physical security of NRC telecommunications systems, and found that improvements were needed to strengthen controls over the use of telecommunications services and the physical security of NRC telecommunications systems.

Open Recommendations	Actions Pending
3. Revise Management Directive and Handbook 2.3 to include effective management controls over NRC Headquarters staff use of agency telecommunications services.	The revised management directive and handbook is in final concurrence and is expected to be issued by January 31, 2008.

AUDIT OF NRC'S DECOMMISSIONING PROGRAM (OIG-05-A-17)   SEPTEMBER 30, 2005

This audit was conducted to determine whether NRC's decommissioning program achieves desired performance results as stated in the Strategic Plan and reported in the Performance and Accountability Report. The audit identified that while NRC's decommissioning program has processes in place to monitor, evaluate, and report on performance, some performance results could not be verified. In addition, the audit found that although most of the recommendations from an FY 2003 self-evaluation of the program were implemented, progress to implement a few was minimal.

Open Recommendations	Actions Pending
1. Clarify and disseminate expectations for generating and maintaining supporting documentation for performance data to staff responsible for preparing and collecting performance data.	Revised Management Directive 4.7, "Planning, Budgeting, and Performance Management," will include clarifications of expectations for generating and maintaining supporting documentation for performance data. The revised management directive and handbook is expected to be issued in February 2008.

SYSTEM EVALUATION OF SECURITY CONTROLS FOR STANDALONE PERSONAL COMPUTERS AND LAPTOPS (OIG-05-A-18)   SEPTEMBER 30, 2005

This evaluation was conducted as part of the OIG's review of NRC's implementation of the Federal Information Security Management Act for FY 2005, with the objectives of evaluating the effectiveness of NRC security policies, procedures, practices, and controls for standalone personal computers (PCs) and laptop computers. The review found that security controls for standalone PCs and laptops were not adequate, that the devices were not monitored for compliance with Federal regulations, and agency information technology coordinators' understanding of disposal practices for these devices were not consistent.

Open Recommendations	Actions Pending
1. Provide users guidance for implementing security controls on standalone PCs and laptops.	By 2010, guidance for implementing security controls on standalone PCs and laptops will be developed and posted on the computer security Web page, and offices will be notified that the guidance is available.
2. Develop and require users to sign a rules-of-behavior agreement accepting responsibility for implementing security controls on standalone PCs and laptops.	Standard rules of behavior implementing security controls on standalone PCs and laptops will be developed, the standard agreement will be posted on the computer security Web page, and offices will be notified of the requirement for all users of such devises to sign the agreement as a condition of using the devices. Development of the rules of behavior, including review by the National Treasury Employees Union, is expected to be completed by the end of FY 2008.
3. Develop and implement procedures for verifying all required security controls are implemented on standalone PCs and laptops.	By 2010, procedures for verifying all required security controls are implemented on standalone PCs and laptops will be developed and implemented.

continued

SYSTEM EVALUATION OF SECURITY CONTROLS FOR STANDALONE PERSONAL COMPUTERS AND LAPTOPS (OIG-05-A-18), CONTINUED

Open Recommendations	Actions Pending
4. Provide users guidance on compliance with Executive Order (EO) 13103, Computer Software Piracy, for standalone PCs and laptops.	Clear guidance on compliance with EO 13103, for standalone PCs and laptops will be developed and disseminated as part of the standard rules of behavior as discussed above under Recommendation 2. Development of the rules of behavior, including review by the National Treasury Employees Union, is expected to be completed by the end of FY 2008.
5. Develop and require users to sign a rules-of-behavior agreement acknowledging their compliance with EO 13103, Computer Software Piracy, for standalone PCs and laptops.	As part of the development of the standard rules of behavior as discussed above under Recommendations 2 and 4, a standard rules-of-behavior agreement for users to acknowledge their compliance with EO 13103 for standalone PCs and laptops will be developed and offices will be notified of the requirement for all users of such devices to sign the agreement as a condition of using the devices. Development of the rules of behavior, including review by the National Treasury Employees Union, is expected to be completed by the end of FY 2008.
6. Develop and implement procedures for monitoring compliance with EO 13103, Computer Software Piracy, for standalone PCs and laptops.	Procedures for monitoring compliance with EO 13103 for standalone PCs and laptops will be developed and issued as part of the as part of the standard rules of behavior as discussed above under Recommendation 2. Development of the rules of behavior, including review by the National Treasury Employees Union, is expected to be completed by the end of FY 2008.
7. Develop detailed procedures in the appropriate NRC management directives (MDs) for the disposal of equipment used to process safeguards and/or classified information. These procedures should then be referenced in the appropriate chapters of the Volume 12 series of management directives.	NRC's process for disposing of media/equipment used to process safeguards and/or classified information at Headquarters and regional offices was documented in January 2007. MD 12.1, NRC Facility Security Program, and MD 12.2, NRC Classified Information Security Program, were revised to include language consistent with guidance currently provided in MD 12.5, NRC Automated Information Security Program, and reissued on August 2, 2007. The appropriate language has also been incorporated into draft new MD 12.7, NRC Safeguards Information Security Program, which is currently in final agency concurrence and is expected to be issued by December 2007.
8. Include the procedures for the disposal of equipment containing safeguards and/or classified information in the security plan templates.	The standard security plans for systems that process safeguards information or classified information have been modified to contain procedures for the disposal of equipment containing such information. Closure of this recommendations requires OIG's verification that the modified security plan templates include the citations to reference the appropriate disposal procedures, which is expected in early FY 2008.

NRC'S GENERIC COMMUNICATIONS PROGRAM (OIG-05-A-19) OCTOBER 7, 2005

This audit was conducted to assess the effectiveness of the Generic Communications Program, specifically whether NRC generic communications are issued in accordance with the Generic Communications Program and other regulatory requirements, and how NRC tracks licensee actions on generic communications. The audit found that NRC has an established framework for developing and issuing certain generic communications, but that weaknesses exist in NRC's internal controls over generic communications in controls for oversight of licensee actions.

Open Recommendations	Actions Pending
1. Include safeguards advisories, as well as any other agency communication tool that meets the definition of a generic communication, in the formal Generic Communications Program to ensure compliance with regulatory requirements.	Proposed new "Management Directive (MD) "8.18," NRC Generic Communications Program," defines the scope of NRC's generic communications and defines organizational roles and responsibilities for each generic communications product, and establishes security advisories and Information Assessment Team advisories as additional agency generic communications products. The revised MD is in final concurrence and is expected to be issued in FY 2008.
3. Implement controls to ensure a systematic, consistent tracking methodology from initiation to closure for each agency-issued generic communication.	In June 2006, NRC established an interoffice working group to evaluate the current process for initiating, developing, tracking, and distributing generic communications and recommend how the process should be changed. The working group decided to incorporate the database into the project plan for a system to track and store requests for additional information (licensee responses and inquiries) which includes capabilities for collaborative discussion threads and for tracking NRC reviewers' queries to the industry and the responses. The initiation phase of the project began in June 2007, and the staff expects to have a prototype database before the end of 2007.
4. Direct the development of a methodology that will allow the staff to gauge the effectiveness of agency-issued generic communications.	Proposed new MD 8.18, "NRC Generic Communications Program," defines the scope of NRC's generic communications and defines organizational roles and responsibilities for each generic communications product, including the conduct of effectiveness reviews. In addition, it clearly identifies those generic communications that require effectiveness reviews. The revised MD is in final concurrence and is expected to be issued in FY 2008.

INDEPENDENT EVALUATION OF NRC'S IMPLEMENTATION OF THE FEDERAL INFORMATION SECURITY MANAGEMENT ACT FOR FY 2005 (OIG-05-A-21)   OCTOBER 7, 2005

This was an independent evaluation of NRC's implementation of the Federal Information Security Management Act for FY 2005. The review found that while NRC had made improvements to its automated information security program, there were major deficiencies that needed to be addressed.

Open Recommendations	Actions Pending
1. Categorize all NRC information systems, including systems operated by a contractor or other organization on behalf of the agency, in accordance with Federal Information Processing Standard (FIPS) 199.	Over half of the NRC's Major Applications, General Support Systems, or contractor systems have an approved security categorization. The remaining systems are either in process. The categorization of the remaining systems in accordance with FIPS 199 is expected to be completed by December 31, 2007. OIG is combining this recommendation with Recommendation 2 of OIG-07-A-19, the Independent Evaluation of NRC's Implementation of FISMA for FY 2007, which will be tracked in the FISMA plan of action and milestones (POA&M) until closed.
3. Develop and implement procedures to ensure contingency plans are tested annually, regardless of the status of the systems' certification and accreditation.	In September 2007, NRC informed OIG that a procedure addressing this requirement was issued on July 1, 2007. Closure of this recommendation requires OIG review of the procedure, which is expected in early FY 2008. OIG is combining this recommendation with Recommendation 6 of OIG-07-A-19, the Independent Evaluation of NRC's Implementation of FISMA for FY 2007, which will be tracked in the FISMA POA&M until closed.
4. Maintain current copies of certification and accreditation (C&A) memoranda for systems provided by other Federal agencies.	NRC has C&A documentation for four of the eight systems provided to NRC by other Federal agencies. The C&A memoranda for the four outstanding systems have not yet been submitted by the responsible system owners, although they are not required to be submitted until an Authority to Operate is requested, which is expected to be completed by December 31, 2007. OIG is combining this recommendation with Recommendation 7 of OIG-07-A-19, the Independent Evaluation of NRC's Implementation of FISMA for FY 2007, which will be tracked in the FISMA POA&M until closed.

continued

INDEPENDENT EVALUATION OF NRC'S IMPLEMENTATION OF THE FEDERAL INFORMATION SECURITY MANAGEMENT ACT FOR FY 2005 (OIG-05-A-21), CONTINUED

Open Recommendations	Actions Pending
5. Maintain current copies of self-assessments for systems provided by other Federal agencies.	To satisfy this recommendation, NRC is to obtain a letter from the senior IT security officer at other Federal agencies who provide systems to NRC stating that the annual self-assessment for the system in question has been completed. This requirement will be communicated by the Senior Information Technology Security Officer (SITSO) to NRC offices on an annual basis to remind responsible offices to update the self-assessment status for their systems. The Office of Information Services (OIS) will keep track of the status of self-assessment for all systems. The self-assessment documentation (memorandum or e-mail) will be reflected as an artifact requirement for "Other Government agency" systems. These actions are expected to be completed by December 31, 2007. OIG is combining this recommendation with Recommendation 7 of OIG-07-A-19, the Independent Evaluation of NRC's Implementation of FISMA for FY 2007, which will be tracked in the FISMA POA&M until closed.
6. Maintain current copies of annual contingency plan testing results for systems provided by other Federal agencies.	To satisfy this recommendation, NRC is to obtain a letter from the senior IT security officer at other Federal agencies who provide systems to NRC stating that the annual contingency plan testing for the system in question has been completed. This requirement will be communicated by the SITSO to NRC offices on an annual basis to remind responsible offices to update the annual contingency plan testing status for their systems. OIS will keep track of the status of annual contingency plan testing for all systems. This annual contingency plan testing documentation (memorandum or e-mail) will be reflected as an artifact requirement for "Other Government agency" systems. These actions are expected to be completed by December 31, 2007. OIG is combining this recommendation with Recommendation 7 of OIG-07-A-19, the Independent Evaluation of NRC's Implementation of FISMA for FY 2007, which will be tracked in the FISMA POA&M until closed.
7. Develop and implement procedures for performing oversight of major applications and general support systems operated by a contractor or other organization on behalf of the agency.	In July 2007, NRC informed OIG that a procedure addressing this requirement was issued on March 1, 2007. Closure of this recommendation requires OIG review of the procedure, which is expected in early FY 2008. OIG is combining this recommendation with Recommendation 8 of OIG-07-A-19, the Independent Evaluation of NRC's Implementation of FISMA for FY 2007, which will be tracked in the FISMA POA&M until closed.

continued

INDEPENDENT EVALUATION OF NRC'S IMPLEMENTATION OF THE FEDERAL INFORMATION SECURITY MANAGEMENT ACT FOR FY 2005 (OIG-05-A-21), CONTINUED

Open Recommendations	Actions Pending
8. Review and update the six completed e-authentication risk assessments to correct inaccuracies and inconsistencies with FIPS 199 security categorizations.	Of the six completed e-authentication risk assessments (ERAs), three were previously verified as having been updated to correct inaccuracies and inconsistencies with FIPS 199 security categorizations. Of the remaining three, one was completed by the system owner to include e-authentication information and was submitted to the SITSO for review in June 2007. Another system is being redesigned into four subsystems, and the new security categorization with the updated design is being developed by the system owner, so the current e-authentication documentation will be replaced with the new one. The last system's security categorization is being revised to incorporate the SITSO's comments. The remaining ERAs are expected to be completed by December 2007. OIG is combining this recommendation with Recommendation 15 of OIG-07-A-19, the Independent Evaluation of NRC's Implementation of FISMA for FY 2007, which will be tracked in the FISMA POA&M until closed.
9. Develop and implement a plan for completing the remaining e-authentication risk assessments.	Currently, 20 active Major Applications have completed ERAs. Two Major Applications do not have an ERA and a third system's security categorizations use a new template that incorporates the ERA. All remaining ERAs are expected to be completed by December 31, 2007. OIG is combining this recommendation with Recommendation 15 of OIG-07-A-19, the Independent Evaluation of NRC's Implementation of FISMA for FY 2007, which will be tracked in the FISMA POA&M until closed.
10. Develop and implement procedures for ensuring employees and contractors with significant IT security responsibilities are identified, receive security awareness and training, and the individual and associated training are readily identifiable.	NRC is in the process of selecting a security Line of Business agency to provide training. It is expected that the training will begin by October 31, 2008. OIG is combining this recommendation with Recommendation 14 of OIG-07-A-19, the Independent Evaluation of NRC's Implementation of FISMA for FY 2007, which will be tracked in the FISMA POA&M until closed.

AUDIT OF NRC'S INTEGRATED PERSONNEL SECURITY SYSTEM (OIG-06-A-06) JANUARY 9, 2006

This audit was conducted to determine if the Integrated Personnel Security System (IPSS) meets its required operational capabilities. It found that while many users report that the system is easier to use than its predecessor systems and provides more functionality, the IPSS does not perform in accordance with required operational capabilities.

Open Recommendations	Actions Pending
4. Review and correct the most recent reinvestigation dates within the IPSS.	A top-to-bottom cleanup effort of every active file to ensure the most recent reinvestigation dates are in the IPSS is under way, with completion expected by November 30, 2007.
7. Perform top-to-bottom cleanup effort of every active file; support this effort with clear written guidance as to what data goes in what field.	Updated IPSS data entry guidance was issued in February 2007 and that guidance is being used to perform the top-to-bottom cleanup effort of every active file that is under way, and is expected to be completed by November 30, 2007.
17. Conduct a cost-benefit analysis to determine whether the agency should continue to develop the IPSS versus replacing the system. As part of the cost-benefit analysis, consider current Federal personnel security requirements.	A contract was awarded in September 2007 to obtain support for Homeland Security Presidential Directive 12 planning and implementation. NRC included in this contract a task to conduct a cost-benefit analysis that will compare the option of continuing to modify IPSS with the option of replacing the entire system. The results of the this cost-benefit analysis are expected to be available by December 31, 2007.

AUDIT OF NRC'S OFFICE OF NUCLEAR SECURITY AND INCIDENT RESPONSE
(OIG-06-A-09)   FEBRUARY 17, 2006

This audit was an independent evaluation of the operations of the Office of Nuclear Security and Incident Response (NSIR), formed in April 2002, specifically, focusing on NSIR's management of emergent work, communications with stakeholders, and implementation of the recommendations from the organizational assessment performed in 2003. The audit found that while NSIR accomplished a great deal since its inception, it needed to focus on refining and formalizing its day-to-day operations to improve its ability to meet its mission.

Open Recommendations	Actions Pending
1. Establish a means of assessing the current workload and prioritizing assignments, including but not limited to emergent work, as they are received, so they can be incorporated into the workload without overextending NSIR's resources.	NSIR implemented a reorganization in 2006 that established an improved span of control and management of the office's workload and developed improved procedures and processes for tracking controlled correspondence. NSIR continually monitors its performance and effectiveness through its operating plan and the performance plans of its managers and staff, and also uses the Performance Budgeting and Performance Management Process to manage planned and unplanned work. NSIR's Work Planning and Management Initiative Group (WPMIG) will complete a revision to office procedure COM-201, "Controlled Correspondence" and a comprehensive business process framework to identify relevant existing office procedures and processes and determine if any new procedures need to be developed. The revised and any new procedures identified are expected to be issued by December 2007.

continued

AUDIT OF NRC'S OFFICE OF NUCLEAR SECURITY AND INCIDENT RESPONSE
(OIG-06-A-09), CONTINUED

Open Recommendation	Action Pending
2. Review the Emergent Work Process to ensure emergent work is accurately documented to assist with workforce and budget decisions.	NSIR has focused its efforts on integrating the new Electronic Document and Action Tracking System (EDATS) into its overall work planning management system, and expects to fully implement this integrated solution after it is accredited and certified by December 2007.
5. Establish and implement a method to measure the level of effective communications.	NSIR is evaluating the utility of the Division of Preparedness and Response's metrics, developed to measure the level of effective communications, to determine whether they should be expanded to the rest of the organization. This effort is expected to be completed in early FY 2008.
6. Assess the recommendations from the 2003 office assessment to determine their applicability and implement those that would benefit NSIR today.	There were two groups of initiatives to address the recommendations from the 2003 office assessment. The Roles and Responsibilities Initiative resulted in completion of and issuance of new Elements and Standards for branch chiefs for the FY 2007 performance appraisal cycle. The roles and responsibilities team also determined that the automated NSIR Functional Directory (available on the NSIR intranet) provides the necessary information to office employees concerning their roles and responsibilities. While not originally part of the roles and responsibilities team's mandate, the team also reviewed the job responsibilities of Technical Assistants and Management Analysts within NSIR to determine how best to align these positions and associated responsibilities in the organization. This review included comparison of current duties versus position descriptions (PDs) and comparisons across divisions and with other offices. The recommendations from this review were discussed at an NSIR management retreat mid-June 2007, and as a result, NSIR standardized the PDs of the technical assistants and management assistants across the office. NSIR has also made significant progress in providing its employees with copies of their official PDs. About 65 percent of NSIR's PDs have been classified and copies provided to the employee and manager. All PDs are expected to be classified by December 2007.
	As part of the Staffing and Budget Development Initiative, NSIR deployed a staffing plan and vacancy report in March 2007 and issued a training procedure in May 2007.
	Recommendations remaining from the 2003 office assessment relate to IT infrastructure needs. Assessment of these has been delayed until the consolidation of the NSIR staff on two adjacent floors, which is expected to be completed by the end of 2007. It is expected that actions to address those recommendations will be completed in early FY 2008.

AUDIT OF THE DEVELOPMENT OF THE NATIONAL SOURCE TRACKING SYSTEM
(OIG-06-A-10)   FEBRUARY 24, 2006

This audit was conducted to determine whether NRC's oversight of byproduct and sealed source materials provides reasonable assurance that licensees are using the materials safely and account for and control the materials. It concluded that the NSTS may be inadequate because the supporting regulatory analysis is based on unreliable data and did not consider other viable options.

Open Recommendation	Action Pending
1. Before the National Source Tracking System (NSTS) rulemaking is finalized, conduct a comprehensive regulatory analysis for the NSTS that explores other viable options, such as those in the International Atomic Energy agency's (IAEA's) Code of Conduct. The regulatory analysis should include an assessment of expanding materials tracked in NSTS to contain categories 2, 3, 4, and 5; aggregation of sources; and bulk material.	The Commission has directed the NRC staff to prepare a proposed rule to include IAEA Category 3 data in the NSTS. As part of the proposed rulemaking, the staff is developing a technical basis and a regulatory impact analysis to provide the rationale for considering inclusion of licensees with Category 3 and 3.5 sources in the NSTS. In preparing the technical basis and regulatory analysis, the staff will use, as partial input, results and information on the numbers of licensees and sources obtained from a one-time data collection and analysis of Category 3.5 sources, which is under way. A Rulemaking Working Group, consisting of NRC Headquarters technical and legal staff, regional staff, and Agreement State representatives, has been formed to consider technical information and issues associated with including IAEA Category 3 sources in the NSTS. The proposed rule package is scheduled to be submitted to the Commission for review and approval in March 2008. The proposed rule package will include the *Federal Register* notice containing the proposed rule and the draft regulatory impact analysis, and will be issued for public comment upon the Commission's approval.

AUDIT OF THE BYPRODUCT MATERIALS LICENSE APPLICATION & REVIEW PROCESS
(OIG-06-A-11)   MARCH 14, 2006

As part of a larger effort to determine whether NRC's oversight of byproduct material provides reasonable assurance that licensees account for and control the materials, this audit was specifically directed towards determining if NRC ensures, through its license application and review process, that only legitimate entities receive NRC byproduct material licenses. It concluded that because NRC has as not conducted vulnerability assessments of all aspects of the materials program, there may be vulnerabilities in the license application and review process that could be exploited by individuals with malevolent intent.

Open Recommendations	Actions Pending
1. Conduct a complete vulnerability assessment of the materials program, including the license application and review process and the methods used by licensees to purchase byproduct material from sellers.	In September 2007, the Commission approved a comprehensive plan to address needed changes in NRC's process for issuing licenses for radioactive sources. The plan calls for an independent, external review to identify potential weaknesses or security gaps in the NRC materials licensing process, and is expected to be completed by March 2008. Additionally, the plan establishes a Materials Program Working Group that will submit a comprehensive report to September 30, 2008, providing recommendations to address any identified security gaps or weaknesses.

continued

AUDIT OF THE BYPRODUCT MATERIALS LICENSE APPLICATION & REVIEW PROCESS
(OIG-06-A-11) CONTINUED

Open Recommendations	Actions Pending
2. Modify the license application and review process to mitigate the risks identified in the vulnerability assessment.	An initial schedule for completion of recommendations resulting from the independent, external review panel's assessment is expected to be available in early 2008. The independent, external review panel's report will be provided to the Materials Program Working Group for further assessment of the panel's recommendations (i.e., establishing time lines for execution of the recommendations). Depending on the depth and scope of the working group's final assessment, changes to the licensing process could take several years; however, both the panel and the working group are also exploring short-term, interim options that would address any identified vulnerabilities in the shortest amount of time.

AUDIT OF NRC'S IMPLEMENTATION OF HOMELAND SECURITY PRESIDENTIAL DIRECTIVE-12
(OIG-06-A-20)   AUGUST 1, 2006

This audit was conducted to determine whether to determine whether NRC is positioned to meet the requirements of HSPD-12. It found that NRC has implemented a personal identity verification process in accordance with the Office of Management and Budget's deadline and is considering personal identify verification systems that will provide technical interoperability among Government departments and agencies, improvements are needed.

Open Recommendation	Action Pending
6. Formalize the Homeland Security Presidential Directive 12 (HSPD-12) Working Group by developing a charter that defines the membership and expectations.	In September 2007, NRC finalized the HSPD-12 Working Group charter, which defines working group membership and expectations. Closure of this recommendation requires OIG's review of the final charter, which is expected in early FY 2008.

NRC'S BASELINE SECURITY AND SAFEGUARDS INSPECTION PROGRAM
(OIG-06-A-21) SEPTEMBER 8, 2006

The audit of NRC's Drug Testing Program (discussed further in the table on OIG-05-A-05) found that the NRC's Drug-Free Workplace Plan was not in compliance with Federal guidance that requires the plan to receive U.S. Department of Health and Human Services' (HHS's) approval and that it was missing a required clause.

Open Recommendations	Actions Pending
1. Provide the required initial and refresher security training courses for regional security inspectors at the frequency needed to support qualification requirements.	Phase 1 of NRC's corrective actions is to develop foundation security courses, "Security Fundamentals" and "Reactor Technology for Security." The Security Fundamentals course is under review with expected delivery in FY 2008. A pilot for the Reactor Technology for Security course was completed in June 2007 and is under review based on comments received from course participants and lessons learned, with expected delivery in FY 2008. A 3-day Annual Security Refresher Course for security inspectors from all four NRC regions was conducted in November 2006, and is scheduled for November 13-15, 2007. This course is now listed in the NRC course catalog. Phase 2 of NRC's corrective actions is to develop four modules of advanced security field courses. These are being reviewed and NRC is pursuing contracts with outside Federal agencies to provide portions of this specialized training. Phase 2 courses are expected to be available by FY 2009.
4. Update the security inspector training program to ensure course material is current and relevant.	Revisions of the training requirements in NRC Manual Chapter (MC) 1245, Appendix C4, "Safeguards Inspector Technical Proficiency Training and Qualification Journal" and Office of Nuclear Security and Incident Response Office Procedure ADM-109, "Training Development and Qualification Programs" are under development and are expected to be issued in FY 2008 and FY 2009, respectively. As the courses in response to Recommendation 1 are finalized and published in the NRC Training Catalog, MC 1245 and ADM-109 will also be updated.
6. Include guidance in the baseline security and safeguards inspection procedures to ensure inspectors review an adequate number of sample items to assess the effectiveness of the licensee's security program.	The baseline inspection procedures and guidance are currently being reviewed to assess their guidance on sampling. Changes to the baseline inspection procedures to refine the sampling process are expected to be finalized by the end of FY 2008.
7. Implement training on how to select an adequate number of sample items.	The Security Fundamentals course (Module 4, "Security Plans and Requirements"), expected to be delivered in FY 2008, will include a standardized methodology for determining sample sizes. Instruction on the methodology will be included in the Annual Security Refresher course (currently scheduled for November 13-15, 2007) after revisions are made to the baseline inspection procedures.

continued

NRC'S BASELINE SECURITY AND SAFEGUARDS INSPECTION PROGRAM
(OIG-06-A-21), CONTINUED

Open Recommendations	Actions Pending
8. Maintain and share the Office of Nuclear Security and Incident Response database of security findings with the regions.	In August 2006, an updated version of the Security Findings Review Panel (SFRP) database was created. Since its completion, all data from the 4th quarter of calendar year (CY) 2005 to the present has been entered into the database. Based on a review of the database, several upgrades and improvements were recommended and are nearing completion. Data entry of SFRP worksheets prior to the 4th quarter of CY 2005 is complete.

On September 28, 2007, the SFRP database, which provides historic information on security findings, was sent to the NRC Regions. Updates to the database and reports will be disseminated to the regions on a periodic basis. NRC considers action in response to this recommendation to be complete, although closure requires OIG's review of documentation showing that the database is updated and is being sent to the NRC Regions for review, and is expected in early FY 2008. |

AUDIT OF NRC'S PROCESS FOR RELEASING COMMISSION DECISION DOCUMENTS
(OIG-06-A-22) SEPTEMBER 8, 2006

The purpose of this audit was to assess the NRC's process for evaluating SECY Papers and staff requirements memoranda for public release pursuant to relevant legal and regulatory requirements. It concluded that while NRC has a process for handling Freedom of Information Act (FOIA) requests, there are weaknesses in the internal controls needed to ensure full compliance with the FOIA.

Open Recommendations	Actions Pending
1. Develop a program for NRC compliance with the FOIA's automatic disclosure requirements.	Commission procedures have been modified, however, closure of this recommendation requires the revision of Management Directive (MD) 3.4, "Release of Information to the Public," to address how documents will be screened for compliance with 5 U.S.C. 552 (a)(1) and (a)(2). Revised MD 3.4 is expected to be issued by January 2008.
2. Conduct a documented FOIA 552(a)(1) and (a)(2) review of previously unpublished SECY Papers and staff requirements memoranda.	The NRC has disagreed with this recommendation and provided a justification to the OIG in March 2007. The recommendation remains in an unresolved status pending the OIG's additional analysis and consideration of the NRC's justification.

EVALUATION OF NRC'S USE OF PROBABILISTIC RISK ASSESSMENT IN REGULATING THE COMMERCIAL NUCLEAR POWER INDUSTRY (OIG-06-A-24)   SEPTEMBER 29, 2006

The objectives of this evaluation were to determine if NRC is following prevailing good practices in probabilistic risk assessment (PRA) methods and data in its use of PRA, using prevailing good practices in PRA methods and data appropriately in its regulation of nuclear power plant licensees, and achieving the objectives of the PRA policy statement. It concluded that although NRC is employing prevailing good practices in regulation of nuclear power plants, NRC lacks formal, documented processes and associated configuration control for PRA computer models and software.

Open Recommendations	Actions Pending
1. Develop and implement a formal, written process for maintaining PRA models that are sufficiently representative of the as-built, as-operated plant to support model uses.	The NRC's revised Risk Assessment of Operational Events Handbook, was completed in September 2007, now provides a formal, written process for maintaining PRA models to ensure that the Standardized Plant Analysis Risk (SPAR) models used in the risk analysis of operational events represent the as-built, as-operated plant to the extent needed to support the analyses. The revised handbook is expected to be available for implementation in early FY 2008.
3. Conduct a full verification and validation (V&V) of the Systems Analysis Program for Hands-On Integrated Reliability Evaluations (SAPHIRE) Version 7.2 and Graphical Evaluation Module (GEM). (SAPHIRE and GEM are software programs used to perform evaluations of SPAR models and provide risk results based on the events or initiators being evaluated.)	Because development of SAPHIRE Version 8 is under way, a full V&V of SAPHIRE Version 7 would not be an effective use of resources. Therefore, closure of this recommendation requires the general release of SAPHIRE Version 8, which is expected to occur in July 2009.

INDEPENDENT EVALUATION OF NRC'S IMPLEMENTATION OF THE FEDERAL INFORMATION SECURITY MANAGEMENT ACT FOR FY 2006 (OIG-06-A-26)   SEPTEMBER 29, 2006

This was an independent evaluation of NRC's implementation of the Federal Information Security Management Act for FY 2006. The review found that NRC's information security program has various information security program deficiencies and weaknesses.

Open Recommendation	Action Pending
2. Re-categorize the Network Continuity of Operations (COOP) listed system as a general support system.	NRC has incorporated the components of the COOP system into existing Infrastructure General Support Systems, and updated the security categorization documents for the Local Area Network/Wide Area Network (LAN/WAN), e-mail, Remote Access System (RAS), and Novell Infrastructure systems to incorporate the appropriate COOP components. The updates to the security categorization documents have been completed and approved for the LAN/WAN and e-mail systems, and the updates to the security categorization documents for the RAS and Novell Infrastructure systems were completed and forwarded to the Senior Information Technology Security Officer (SITSO) for approval. Final approval of the security categorization documents for the RAS and Novell Infrastructure systems is expected to be given by December 15, 2007. OIG is combining this recommendation with Recommendation 9 of OIG-07-A-19, the Independent Evaluation of NRC's Implementation of FISMA for FY 2007, which will be tracked in the FISMA plan of action and milestones (POA&M) until closed.

SUMMARY OF FINANCIAL STATEMENT AUDIT AND MANAGEMENT ASSURANCES

SUMMARY OF FINANCIAL STATEMENT AUDIT

Audit Opinion — Unqualified

Restatement — No

Material Weaknesses	Beginning Balance	New	Resolved	Consolidated	Ending Balance
Information System-wide Security Controls	1	-	-	-	1
Fee Billing System	1	-	(1)	-	-
Total Material Weaknesses	2	-	(1)	-	1

SUMMARY OF MANAGEMENT ASSURANCES

Effectiveness of Internal Control over Financial Reporting (FMFIA § 2)

Statement of Assurance — Unqualified

There are no material weaknesses for Internal Control over Financial Reporting.

Effectiveness of Internal Control over Operations (FMFIA § 2)

Statement of Assurance — Qualified

Material Weaknesses	Beginning Balance	New	Resolved	Consolidated	Reassessed	Ending Balance
Information System-wide Security Controls	1	-	-	-	-	1
Information System-wide Security Controls	1	-	-	(1)	-	-
Total Material Weaknesses	2	-	-	(1)	-	1

Conformance with Financial Management System Requirements (FMFIA § 4)

Statement of Assurance — Systems conform to financial management system requirements

Non-Conformances	Beginning Balance	New	Resolved	Consolidated	Reassessed	Ending Balance
Fee Billing System	1	-	(1)	-	-	-
Total Non-Conformances	1	-	(1)	-	-	-

Compliance with Federal Financial Management Improvement Act (FFMIA)

	Agency	Auditor
Overall Substantial Compliance	No	No
1. Systems Requirements	No	No
2. Accounting Standards	Yes	Yes
3. USSGL at Transaction Level	Yes	Yes

VERIFICATION AND VALIDATION OF NRC'S MEASURES AND METRICS

Below is the agency's verification and validation of the measures and metrics associated with the agency's strategic goals of Safety and Security.

NRC Data Collection Procedures

Most of the data used to measure the NRC's performance against its strategic goals related to safety are obtained or derived from the NRC's abnormal occurrence (AO) data and reports submitted by licensees. The AO criteria have been amended to ensure that they are consistent with the NRC's rulemaking on Title 10, Part 35, "Medical Use of Byproduct Material," of the *Code of Federal Regulations* (10 CFR Part 35).

The NRC developed its AO criteria to comply with the legislative intent of section 208 of the Energy Reorganization Act of 1974, as amended. This act requires the NRC to inform the Congress of unscheduled incidents or events that the Commission determines to be significant from the standpoint of public health and safety. The agency includes events that meet the AO criteria in its annual "Report to Congress on Abnormal Occurrences" (NUREG-0090). In addition, in 1997, the Commission determined that events occurring at Agreement State-licensed facilities that meet the AO criteria should be reported in the annual AO report to Congress. Therefore, the AO criteria developed by the NRC are uniformly applied to events that occur at facilities, licensed or otherwise, that are regulated by the NRC and the Agreement States.

Data for AOs originate from external sources, such as Agreement States and NRC licensees. The NRC believes that these data are credible because (1) NRC regulations require the reporting of the information needed from external sources; (2) the NRC maintains an aggressive inspection program that, among other activities, audits licensees and evaluates Agreement State programs to determine whether information is being reported as required by the regulations; and (3) agency procedures address reviewing and evaluating licensees. The NRC database systems that support this process include the Licensee Event Report Search (LERSearch) system, the Accident Sequence Precursor (ASP) database, the Nuclear Materials Events Database (NMED), and the Radiation Exposure Information Report system.

The NRC has established procedures for the systematic review and evaluation of events reported by NRC licensees and Agreement State licensees. The objective of the review is to identify events that are significant from the standpoint of public health and safety based on criteria that include specific thresholds. The NRC uses a number of sources to determine the reliability and technical accuracy of event information reported to the agency. Such sources include (1) the NRC licensee reports, which are carefully analyzed, (2) NRC inspection reports, (3) Agreement State reports, (4) periodic reviews of Agreement State regulatory programs, (5) NRC consultant/contractor reports, and (6) U.S. Department of Energy operating experience weekly summaries. In addition, daily interactions and exchanges of event information occur between headquarters and the regional offices, and staff participate in periodic conference calls among headquarters, the regions, and Agreement States to discuss event information. All applicable NRC Headquarters program offices, regional offices, and agency management personnel validate and verify identified events that meet the AO criteria before their submission to Congress.

The Agency Action Review meeting provides another opportunity for NRC's senior management to discuss significant events, licensee performance issues, trends, and actions that the NRC needs to take to mitigate recurrences.

The agency's computer security program maintains data protection and provides administrative, technical, and physical security measures to guard the agency's information, automated information systems, and information technology infrastructure. These measures include special safeguards to protect classified information, unclassified safeguards information, and sensitive unclassified information that are processed, stored, or produced on designated automated information systems.

GOAL 1   SAFETY

Ensure protection of public health and safety and the environment.

Nuclear Reactor Safety

Strategic Outcomes:

• No nuclear reactor accidents.

• No inadvertent criticality events.

• No acute radiation exposures resulting in fatalities.

• No releases of radioactive materials that result in significant radiation exposures.

• No releases of radioactive materials that cause significant adverse environmental impacts.

Verification:

Licensees report any nuclear reactor events at their facilities in licensee event reports (LERs). The NRC reviews the LER data, and the agency's AO coordinators then discuss each potential AO during their periodic meetings at headquarters and the regional offices to determine whether it meets the AO reporting criteria. The staff use the LERs to identify any nuclear reactor accidents, deaths from acute radiation exposures, events that result in significant radiation exposure, or releases of radioactive materials that cause significant adverse environmental impacts that meet the criterion for an AO. In addition, NRC specialists periodically conduct inspections to assess licensee compliance with reporting criteria as well as

radiological and environmental release criteria. If a licensee reports an event involving core damage, NRC inspectors carefully investigate the event to ensure the validity of the information in the licensee's report. In addition, a resident inspector on duty at each reactor monitors the facility in real time. The resident inspector verifies the safe operation of the facility and would be aware of any instances in which core damage has occurred or radiation was released from the reactor in excess of reporting limits.

The NRC staff prepares AO write-ups and evaluates events, using specific criteria to select those events that the staff recommends to the Commission to be considered as AOs. The NRC's Office of Nuclear Regulatory Research makes the final determination about which events to recommend for consideration as potential AOs. NRC Management Directive 8.1, "Abnormal Occurrence Reporting Procedure," provides thorough documentation of the AO reporting process.

Validation: Validation addresses the issues below.

No nuclear reactor accidents. The NRC Severe Accident Policy Statement defines nuclear reactor accidents as those events that result in substantial damage to the reactor fuel, regardless of whether offsite consequences occur.

No inadvertent criticality events. Events collected under this performance measure are actual occurrences of accidental criticality. Such events could compromise public health and safety, the environment, and the common defense and security. Events of this magnitude are not expected and would be rare. If such an event occurs, it would result in a prompt and thorough investigation, including consequences, root causes, and necessary actions by the licensee and the NRC to mitigate the consequences and prevent recurrence.

No acute radiation exposures resulting in fatalities. Determining whether any deaths result from acute radiation exposure is fundamentally essential to protecting public health and safety. Events of this magnitude are rare. If such an unlikely event occurs, it would result in a prompt and thorough investigation of the event, its consequences, its root causes, and necessary actions by the licensee and/or the NRC to mitigate the consequences and prevent recurrence. This strategic outcome measure is a direct measurement of the occurrence of radiation-related deaths at nuclear reactors.

No releases of radioactive materials that result in significant radiation exposures. Nuclear power generation produces radiation, which can be harmful if not properly controlled. Measuring the number of events resulting in significant radiation exposures, as well as any deaths from radiation exposure, indicates whether radiation-related deaths and illness are being prevented. Significant radiation exposures are defined as those that result in unintended permanent functional damage to an organ or a physiological system, as determined by a physician in accordance with AO Criterion I.A.3.

No releases of radioactive materials that cause significant adverse environmental impacts. The radiation produced in the process of generating power from nuclear materials can also potentially harm the environment if it is not properly controlled. Releases that have the potential to adversely impact the environment are currently undefined. As a surrogate for this performance measure, the NRC collects data on the frequency of radiation releases into the environment that exceed specified limits. AO Criterion I.B.1 in Appendix A to NUREG-0090 defines such releases as those involving "the release of radioactive material to an unrestricted area in concentrations which, if averaged over a period of 24 hours, exceed 5,000 times the values specified in Table 2 of Appendix B to 10 CFR Part 20, unless the licensee has demonstrated compliance with 20.1301 using 20.1302(b)(1) or 20.1302 (b)(2)(ii)." The essence of the criterion is that events that result in unintended permanent functional damage to an organ or a

physiological system as determined by a physician are used as the measure for events that result in releases of radioactive material causing an adverse impact on the environment. Such events are reported in LERs, which are sent to the NRC as documentation of reportable occurrences. This strategic outcome measure is a direct measurement of instances in which harmful impacts on the environment occur because of nuclear reactors.

Performance Measures:

- number of significant safety events and conditions per year at reactor facilities

- number of new conditions evaluated as red by the NRC's Reactor Oversight Process, with a reactor safety target of less than or equal to 3[1]

Verification:

The data for this performance measure is collected in two ways as part of the NRC's reactor oversight process (ROP). Inspection findings are collected at least quarterly by NRC inspectors. Inspectors use formal detailed inspection procedures to review plant operations and maintenance. Inspection findings are reviewed by NRC managers to assess their significance as part of the ROP's significance determination process. The data for performance indicators is collected by licensees and submitted to the NRC at least quarterly. The significance of the data is determined by thresholds for each indicator. The NRC conducts inspections of licensees' processes for collecting and submitting the data to ensure completeness, accuracy, consistency, timeliness, and validity.

The NRC enhances the quality of its inspections through inspector feedback and periodic reviews of results, and inspectors are trained through a rigorous qualification program. The quality of performance indicators is improved through continuous feedback

---

[1] This measure is the number of new red inspection findings during the fiscal year plus the number of new red performance indicators during the fiscal year. Programmatic issues at multi-unit sites that result in red findings for each individual unit are considered separate conditions for purposes of reporting for this measure. A red performance indicator and a red inspection finding that are due to an issue with the same underlying causes are also considered separate conditions for purposes of reporting for this measure. Red inspection findings are included in the fiscal year in which the final significance determination was made. Red performance indicators are included in the fiscal year in which Reactor Oversight Process external Web page was updated to show the red indicator.

from licensees and inspectors that is incorporate into guidance documents. The NRC publishes the inspection findings and performance indicators on the agency's Web site, and incorporates feedback received from all stakeholders as appropriate.

Validation:

The inspection findings and performance indicators used by the ROP cover a broad range of plant operations and maintenance. NRC managers review significant issues that are identified, and inspectors conduct supplemental inspections of selected aspects of plant operations as appropriate. Senior agency managers annually review plants that are identified as having performance issues, as well as a self-assessment of the ROP, and then report the results to the Commission.

This measure indicates the number of new red inspection findings during the fiscal year plus the number of new red performance indicators during the fiscal year. Programmatic issues at multiunit sites that result in red findings for each individual unit are considered as separate conditions for purposes of reporting for this measure. A red performance indicator and a red inspection finding that are attributable to an issue with the same underlying causes are also considered as separate conditions for purposes of reporting for this measure. Red inspection findings are included in the fiscal year in which the final significance determination was made. Red performance indicators are included in the fiscal year in which the ROP external Web page was updated to show the red indicator.

- number of significant safety events and conditions per year at reactor facilities

- number of significant ASPs of a nuclear accident, with a reactor safety target of 0[2]

Verification:

The Commission has an ASP program to systematically evaluate U.S. nuclear power plant operating experience to identify, document, and rank those operating events that were most significant in terms of the potential for inadequate core cooling and core damage (i.e., precursors). The ASP program evaluation process has five steps. First, the NRC screens operating experience data to identify events and/or conditions that may be potential precursors to a nuclear accident. The data that are evaluated include LERs from the LERSearch database, incident investigation team or augmented inspection team reviews, the NRC's daily screening of operational events, and other events identified by NRC staff as candidates. Second, the staff conducts an engineering review of these screened events using specific criteria to identify those events requiring detailed analyses as candidate precursors. Third, the NRC staff calculates a conditional core damage probability by mapping failures observed during the event to accident sequences in risk models. Fourth, the preliminary potential precursor analyses are provided to the NRC staff and the licensee for independent peer review. However, for ASP analyses of non-controversial, low-risk precursors for which the ASP results reasonably agree with the significance determination process results, licensees may not perform formal peer reviews. The NRC staff will continue to perform an in-house review process for all analyses. Fifth, the NRC provides findings from the analyses to the licensee and the public.

It must also be noted that a time lag exists in obtaining ASP analysis results because they are often based on LERs (submitted up to 60 days after an event), and most analyses take approximately 6 months to finalize. The agency will report final data in the year in which the event occurred.

---

[2] Significant Accident Sequence Precursor (ASP) events have a conditional core damage probability (CCDP) or ΔCDP of $> 1 \times 10^{-3}$. Such events have a 1/1000 (10^{-3}) or greater probability of leading to a reactor accident involving core damage. An identical condition affecting more than one plant is counted as a single ASP event if a single accident initiator would have resulted in a single reactor accident. One event was identified in FY 2002 as having the potential of being a significant precursor. This precursor involved reactor pressure vessel head degradation at Davis-Besse. The detailed ASP Program preliminary analysis of this complex event was completed in September 2004. Based on the screening and engineering evaluation of FY 2002, FY 2003, and FY 2004 events, no other potentially significant precursor were identified. Therefore, the second performance measure was not exceeded for FY 2002, FY 2003, and FY 2004.

Validation:

The ASP program identifies significant precursors as those events that have a 1/1000 (10^{-3}) or greater probability of leading to a nuclear reactor accident. Significant ASP events have a conditional core damage probability or ΔCDP of greater than or equal to 1×10^{-3}.

- number of operating reactors whose integrated performance entered the Inspection Manual Chapter 0350 process, the multiple/repetitive degraded cornerstone column, or the unacceptable performance column of the ROP Action Matrix, with a reactor safety target of less than or equal to 4[3]

Verification:

The NRC's ROP collects data for this performance measure continuously, and the agency publishes the information at least quarterly. NRC inspectors use detailed formal procedures to inspect licensee performance, and NRC managers review the results to ensure the completeness, accuracy, consistency, timeliness, and validity of the data.

The NRC enhances the quality of its inspections through inspector feedback and periodic reviews of results, and inspectors are trained through a rigorous qualification program. The quality is also improved through continuous feedback from licensees and inspectors that is incorporated into guidance documents. The NRC publishes the data on the agency's Web site and incorporates feedback received from all stakeholders as appropriate.

Validation:

The information collected by the ROP covers a broad range of plant operations and maintenance. NRC managers review significant issues that are identified,

and inspectors conduct supplemental inspections of selected aspects of plant operations as appropriate. Senior managers annually review plants that are identified as having performance issues, as well as the agency's self-assessment of the ROP, and then report the results to the Commission.

This measure is the number of plants that have entered the Inspection Manual Chapter 0350 process, the multiple/repetitive degraded cornerstone column, or the unacceptable performance column during the fiscal year (i.e., were not in these columns or process the previous fiscal year). Data for this measure are obtained from the NRC external Web Action Matrix summary page, which provides a matrix of the five columns with the plants listed within their applicable columns and notes the plants in the Inspection Manual Chapter 0350 process. For reporting purposes, plants that are the subject of an approved deviation from the Action Matrix are included in the column or process in which they appear on the Web page.

- number of significant adverse trends in industry safety performance, with a reactor safety target of less than or equal to 1[4]

Verification:

The data for this performance measure are derived from data supplied by all power plant licensees in LERs, data from monthly operating reports, and performance indicator data submitted for the ROP. These data are (1) required by 10 CFR 50.73, "License Event Report System," and/or plant-specific technical specifications or (2) submitted by all plants as part of the ROP. Detailed NRC guidelines and procedures are in place to control each of these reporting processes. The NRC reviews these procedures for appropriateness both periodically and in response to licensee feedback. The NRC also conducts periodic

---

[3] This measure is the number of plants that have entered the Manual Chapter 0350 process, the multiple/repetitive degraded cornerstone column, or the unacceptable performance column during the fiscal year (i.e., were not in these columns or process the previous fiscal year). Data for this measure is obtained from the NRC external web Action Matrix Summary page, that provides a matrix of the five columns with the plants listed within their applicable column and notes the plants in the Manual Chapter 0350 process. For reporting purposes, plants that are the subject of an approved deviation from the Action Matrix are included in the column or process in which they appear on the web page. The target value is set based on the expected addition of several indicators and a change in the long-term trending methodology (which will no longer be influenced by the earlier data and will be more sensitive to changes in current performance).

[4] Considering all indicators qualified for use in reporting.

inspections of licensee processes for collecting and submitting the data to ensure completeness, accuracy, consistency, timeliness, and validity.

All licensees report the data at least quarterly. The NRC staff reviews all of the data and conducts inspections to verify safety-significant information. The NRC also employs a contractor to review the data submitted by licensees, enter the data in a database, and compile the data into various indicators. Quality assurance processes for this work have been established and included in the statement of work for the contract. Administration of the contract controls the experience and training of key personnel. The contractor identifies discrepancies and submits them to both licensees and the NRC for resolution. The NRC reviews the indicators and publishes them on the agency's Web site quarterly. The agency also incorporates feedback from licensees and the public as appropriate.

The target value is based on the expected addition of several indicators and a change in the long-term trending methodology (which will no longer be influenced by the earlier data and will be more sensitive to changes in current performance).

Validation:

The data and indicators that support reporting against this performance measure provide a broad range of information on nuclear power plant performance. The NRC staff tracks indicators and applies statistical techniques to obtain an indication of whether industry performance is improving, steady, or degrading over time. If the staff identifies any adverse trends, the NRC addresses the problem through its processes for handling generic safety issues and issuing generic communications to licensees. The NRC is developing additional risk-informed indicators to enhance the current set of indicators. In doing so, the staff considers the costs and benefits of collecting the data through ongoing, extensive interactions with industry regarding the indicators.

Senior managers annually review the Industry Trends program and report the results to the Commission.

- number of events with radiation exposures to the public and occupational workers from nuclear reactors that exceed AO Criterion I.A with a reactor safety target of 0

Verification:

Licensees report overexposures through the Sequence Coding and Search System (SCSS) LER database, maintained at the Oak Ridge National Laboratory, which receives all LERs and codes them into a searchable database. The SCSS database is used to identify those LERs that report overexposures. NRC resident inspectors stationed at each nuclear power plant provide a high degree of assurance that all events meeting reporting criteria are reported to the NRC. In addition, the NRC conducts inspections if there is any indication that an exposure exceeded or could have exceeded a regulatory limit. Moreover, areas of the facility that may be subject to radiation contamination have monitors that record radiation levels. These monitors would immediately reveal any instances in which high levels of radiation exposure occurred.

Validation:

Given the nature of the process of using radioactive materials to generate power, overexposure to radiation is a potential danger from the operation of nuclear power plants. Such exposure to radiation that exceeds the applicable regulatory limits may potentially occur through either a nuclear accident or other malfunctions at the plant. Consequently, tracking the number of overexposures that occur at nuclear reactors is an important indicator of the degree to which safety is being maintained.

- number of radiological releases to the environment from nuclear reactors that exceed applicable regulatory limits, with a reactor safety target of less than or equal to 2[5]

---

[5]   Beginning in FY 2005, this measure is based upon AO Criterion I.A. Prior to FY 2005, the criterion was based upon a higher threshold of significant functional damage to organs or physiological systems. Using the pre-FY 2005 criteria, NRC reported zero events through FY 2004. However, it should be noted that if the FY 2005 performance measure, based upon AO Criterion I.A. had been in place in FY 2003, two materials events would have been reported for that fiscal year.

Verification:

As with worker overexposures, licensees report environmental releases of radioactive materials that exceed regulations or license conditions through the SCSS LER database maintained at the Oak Ridge National Laboratory. The SCSS database will be used to identify those LERs reporting releases, and the number of reported releases is then applied to this measure. The NRC also conducts periodic inspections of licensees to ensure that they properly monitor and control releases to the environment through effluent pathways. In addition, onsite monitors would record any instances in which the plant releases radiation into the environment. If the inspections or the monitors reveal any indication that an accident or inadvertent release has occurred, the NRC conducts follow-up inspections.

Validation:

The generation of nuclear power creates radioactive materials that are released into the environment in a controlled manner. These radioactive discharges are subject to regulatory controls that limit the quantity discharged and the resultant dose to members of the public. Consequently, the NRC tracks all releases of radioactive materials in excess of regulatory limits as a performance measure because large releases that exceed regulatory limits have the potential to endanger public safety or harm the environment. The NRC inspects every nuclear power plant for compliance with regulatory requirements and specific license conditions related to radiological effluent releases. The inspection program includes enforcement actions to be taken for violations of the regulations or license conditions, based on the severity of the event.

This performance measure includes dose values that are classified as being as low as reasonably achievable (ALARA), as defined in Appendix I, "Numerical Guides for Design Objectives and Limiting Conditions for Operation to Meet the Criterion 'As Low As Is Reasonably Achievable' For Radioactive Material in Light-Water-Cooled Nuclear Power

Reactor Effluents," to 10 CFR Part 50, "Domestic Licensing of Production and Utilization Facilities," as well as the public dose limits in 10 CFR Part 20, "Standards for Protection Against Radiation." Because the performance measure includes ALARA values, which are not safety limits, and because Appendix I to 10 CFR Part 50 allows licensees to temporarily exceed, for good reason, the ALARA dose values, the performance measure is 2.

GOAL 1   SAFETY

Ensure protection of public health and safety and the environment.

Nuclear Material and Waste Safety

Strategic Outcomes:

- No inadvertent criticality events
- No acute radiation exposures resulting in fatalities
- No releases of radioactive materials that result in significant radiation exposures
- No releases of radioactive materials that cause significant adverse environmental impacts

Verification: Verification addresses the issues discussed below.

No inadvertent criticality events. Inadvertent criticality events must be reported, regardless of whether they result in exposures or injuries to workers or the public and regardless of whether they result in adverse impacts to the environment. Licensees immediately report criticality events to the NRC Headquarters Operations Center by telephone through the cognizant licensee safety officer. Follow-up written reports must be submitted to the NRC within 30 days of the initial report. Such reports must contain specific information concerning the event, as specified by 10 CFR 70.50(c)(2) and 10 CFR 76.120(d)(2). The NRC then dispatches an inspection team to confirm the reliability of the data. The event is also tracked through NMED. The NRC would immediately investigate and follow up on an event of this nature.

If an event meeting this threshold occurs, it would be reported to the NRC through a number of sources, but primarily through required licensee notifications. Event notifications and preliminary notifications, which are used to widely disseminate the information to internal and external stakeholders, summarize these events. For activities of the Office of Nuclear Material Safety and Safeguards (NMSS) and the Office of Federal and State Materials and Environmental Management Programs (FSME), NMED is an essential system used to collect information on such events.

The fuel cycle, materials, high-level waste repository, and spent fuel storage and transportation inspection programs are key elements in verifying the completeness and accuracy of licensee reports. The Integrated Materials Performance Evaluation Program (IMPEP) also provides a mechanism to verify that NRC regions are consistently and properly collecting and reporting such events as received from the licensees and entering them in NMED.

The NRC has taken a number of steps to improve the timeliness and completeness of materials event data. These steps include assessment of NMED data during monthly staff reviews; emphasis and analysis during the IMPEP reviews; NMED training in headquarters, the regions, and Agreement States; and discussions at all Agreement State and Conference of Radiation Control Program Directors (CRCPD) meetings.

Validation:

Events collected under this strategic outcome are actual occurrences of accidental criticality. Such events could compromise public health and safety, the environment, and the common defense and security. Events of this magnitude are not expected and would be rare. If such an event occurs, it would result in a prompt and thorough investigation of its consequences, its root causes, and the necessary

actions by the licensee and the NRC to mitigate the situation and prevent recurrence. Therefore, the strategic outcome of no inadvertent criticalities represents a valid measure of ensuring adequate protection of public health and safety.

In assessing the validity of the data collected as appropriate for the strategic outcome, the staff has determined that a logical relationship exists between the data collected and the strategic outcome. Given the magnitude and rarity of a criticality event, the NRC believes that the probability of being unaware of an inadvertent criticality is very small.

Verification: Verification addresses the issues discussed below.

No acute radiation exposures resulting in fatalities. Determining whether a death resulted from acute radiation exposure is fundamentally essential to ensure the protection of public health and safety.

If an event meeting this threshold occurs, it would be reported to the NRC and/or Agreement States through a number of sources, but primarily through required licensee notifications. Event notifications and preliminary notifications, which are used to widely disseminate the information to internal and external stakeholders, summarize these events. For activities of NMSS and FSME, NMED is an essential system used to collect information on such events.

The fuel cycle, materials, high-level waste repository, decommissioning, and spent fuel storage and transportation inspection programs are key elements in verifying the completeness and accuracy of licensee reports. The IMPEP also provides a mechanism to verify that Agreement States and NRC regions are consistently collecting and reporting such events as received from the licensees and entering them in NMED.

The NRC has taken a number of steps to improve the timeliness and completeness of materials event data. These steps include assessment of NMED data during monthly staff reviews; emphasis and analysis during the IMPEP reviews; NMED training in headquarters, the regions, and Agreement States; and discussions at all Agreement State and CRCPD meetings.

Validation:

There is a logical basis for using no acute radiation exposures resulting in fatalities as a strategic outcome for ensuring the protection of public health and safety. The NRC's regulatory process—including licensing, inspection, guidance, regulations, and enforcement activities—is sufficient to ensure that no fatalities are attributable to acute radiation exposure.

Events of this magnitude are not expected and would be rare. In the unlikely event that a death occurs, the NRC or Agreement State technical specialists, with input from expert consultants as necessary, decide whether to ascribe the cause of a death to (1) conditions related to acute radiation exposures or (2) exposure to other radioactive hazardous materials (for fuel cycle activities, this extends to other hazardous materials used with, or produced from, licensed material consistent with 10 CFR Part 70, "Domestic Licensing of Special Nuclear Material").

The NRC believes that the data collected to meet this strategic outcome are free from bias. NMSS and FSME do not use statistical sampling of data to determine results. Rather, they review all events data to determine whether the strategic outcome has been met.

Two important data limitations in determining this strategic outcome are the delay time for receiving information and/or the failure of the NRC to become aware of an event that results in a fatality. Although NMSS and FSME procedures and NRC regulations associated with event reporting include specific requirements for timely notifications, a lag time separates the occurrence of an event and the known consequences of that event.

The NRC believes that the probability of being unaware of a fatality attributable to acute radiation exposure is very small. Periodic licensee inspections

and regulatory reporting requirements are sufficient to ensure that an event of this magnitude would become known.

If such an event occurs, it would result in a prompt and thorough investigation of the event, its consequences, its root causes, and the necessary actions by the licensee and the NRC to mitigate the situation and prevent recurrence. In addition to these immediate actions, the NRC holds periodic meetings where staff and management review events that appear to meet this strategic outcome.

Verification: Verification addresses the issues discussed below.

No releases of radioactive materials that result in significant radiation exposures. NMSS and FSME define this strategic outcome as any discharge or dispersal of radioactive materials from the intended place of confinement—or discharge or dispersal of radioactive wastes during storage, transport, or disposal—that causes significant radiation exposures to a member of the public or occupational worker that directly result in unintended permanent functional damage to an organ or physiological system, as determined by a physician in accordance with AO Criterion I.A.3. (This metric does not include exposures from sealed sources. Exposure from sealed sources would fall under the performance measure for number of events with radiation exposures to the public and occupational workers from radioactive material that exceed AO Criterion I.A.)

If an event meeting this threshold occurs, it would be reported to the NRC and/or Agreement States through a number of sources, but primarily through required licensee notifications. Event notifications and preliminary notifications, which are used to widely disseminate the information to internal and external stakeholders, summarize these events. For activities of NMSS and FSME, NMED is an essential system used to collect information on such events.

The fuel cycle, materials, high-level waste repository, decommissioning, and spent fuel storage and transportation inspection programs are key elements in verifying the completeness and accuracy of licensee reports. The IMPEP also provides a mechanism to verify that Agreement States and NRC regions are consistently collecting and reporting such events as received from the licensees and entering them in NMED.

The NRC has taken a number of steps to improve the timeliness and completeness of materials event data. These steps include assessment of NMED data during monthly staff reviews; emphasis and analysis during the IMPEP reviews; NMED training in headquarters, the regions, and Agreement States; and discussions at all Agreement State and CRCPD meetings.

Validation:

There is a logical basis for using a threshold of no releases of radioactive materials that result in significant radiation exposures as a strategic outcome for ensuring the protection of public health and safety. Significant radiation exposures are defined as those that result in unintended permanent functional damage to an organ or a physiological system, as determined by a physician in accordance with AO Criterion I.A.3. The NRC's regulatory process—including licensing, inspection, guidance, regulations, and enforcement activities—is sufficient to ensure that there are no releases of radioactive materials that result in significant radiation exposures.

Events of this magnitude are not expected and would be rare. In the unlikely event that a significant exposure occurs, NRC or Agreement State technical specialists, with input from expert consultants as necessary, decide whether to ascribe the permanent functional damage to (1) conditions related to acute radiation exposures or (2) exposure to other radioactive hazardous materials (for fuel cycle activities, this extends to other hazardous materials used with, or produced from, licensed material consistent with 10 CFR Part 70).

The NRC believes that the data collected to meet this strategic outcome are free from bias. NMSS and FSME do not use statistical sampling of data to determine results. Rather, they review all event data to determine whether the strategic outcome has been met.

Two important data limitations in determining this strategic outcome are the delay time for receiving information and/or the failure of the NRC to become aware of an event that results in significant radiation exposures. Although NMSS and FSME procedures and NRC regulations associated with event reporting include specific requirements for timely notifications, a lag time separates the occurrence of an event and the known consequences of that event.

The NRC believes that the probability of being unaware of an event that results in significant radiation exposures is very small. Periodic licensee inspections and regulatory reporting requirements are sufficient to ensure that an event of this magnitude would become known.

If such an event occurs, it would result in a prompt and thorough investigation of the event, its consequences, its root causes, and the necessary actions by the licensee and the NRC to mitigate the situation and prevent recurrence. In addition to these immediate actions, the NRC holds periodic meetings where staff and management review events that appear to meet this strategic outcome.

Verification: Verification addresses the issues discussed below.

No releases of radioactive materials that cause significant adverse environmental impacts. Releases that have the potential to cause adverse environmental impacts are currently undefined. The NRC will use as a surrogate any discharge or dispersal of radioactive materials from the intended place of confinement—or discharge or dispersal of radioactive wastes during storage, transport, or disposal—that exceeds the limits for reporting AOs in AO Criterion I.B.

If an event meeting this threshold occurs, it would be reported to the NRC and/or Agreement States through a number of sources, but primarily through required licensee notifications. Event notifications and preliminary notifications, which are used to widely disseminate the information to internal and external stakeholders, summarize these events. For NMSS activities, NMED is an essential system used to collect information on such events.

The fuel cycle, materials, high-level waste repository, decommissioning, and spent fuel storage and transportation inspection programs are key elements in verifying the completeness and accuracy of licensee reports. The IMPEP also provides a mechanism to verify that Agreement States and NRC regions are consistently collecting and reporting such events as received from the licensees and entering them in NMED.

The NRC has taken a number of steps to improve the timeliness and completeness of materials event data. These steps include assessment of NMED data during monthly staff reviews; emphasis and analysis during the IMPEP reviews; NMED training in headquarters, the regions, and Agreement States; and discussions at all Agreement State and CRCPD meetings.

Validation:

There is a logical basis for using releases of radioactive materials that cause significant adverse environmental impacts as a strategic outcome for ensuring the protection of the environment. Releases that have the potential to cause adverse environmental impacts are those that exceed the limits for reporting AOs in AO Criterion I.B.1. The NRC's regulatory process—including licensing, inspection, guidance, regulations, and enforcement activities—is sufficient to ensure that there are no releases of radioactive materials that cause significant adverse environmental impacts.

Events of this magnitude are not expected and would be rare. In the unlikely event of a release of radioactive materials (for fuel cycle activities, this extends to other hazardous materials used with, or produced from, licensed material consistent with 10 CFR Part 70), NRC or Agreement State technical specialists, with input from expert consultants as necessary, decide whether the release caused a significant adverse environmental impact.

The NRC believes that the data collected to meet this strategic outcome are free from bias. NMSS and FSME do not look at statistical sampling of data to determine results. Rather, they review all event data to determine whether the strategic outcome has been met.

Two important data limitations in determining this strategic outcome are the delay time for receiving information and/or the failure of the NRC to become aware of an event that causes significant adverse environmental impacts. Although NMSS and FSME procedures and NRC regulations associated with event reporting include specific requirements for timely notifications, a lag time separates the occurrence of an event and the known consequences of that event.

The NRC believes that the probability of being unaware of an event that causes significant adverse environmental impacts is very small. Periodic licensee inspections and regulatory reporting requirements are sufficient to ensure that an event of this magnitude would become known.

If such an event occurs, it would result in a prompt and thorough investigation of the event, its consequences, its root causes, and the necessary actions by the licensee and the NRC to mitigate the situation and prevent recurrence. In addition to these immediate actions, the NRC holds periodic meetings where staff and management review events that appear to meet this strategic outcome.

Performance Measure:

- number of events with radiation exposures to the public and occupational workers from radioactive material that exceed AO Criteria I.A, with a materials safety target of less than or equal to 6 and a waste safety target of 0

Verification:

This performance measure includes any event involving licensed radioactive materials that results in significant radiation exposures to members of the public and/or occupational workers that exceed the dose limits in the AO reporting criteria. Because of the extremely high doses employed during medical applications of radioactive materials, it is also appropriate to use a radiation exposure that results in unintended permanent functional damage to an organ or a physiological system (as determined by a physician) as a criterion for this measure. AO Criterion I.A is the basis for this measure.

If an event meeting this threshold occurs, it would be reported to the NRC and/or Agreement States through a number of sources, but primarily through required licensee notifications. Event notifications and preliminary notifications, which are used to widely disseminate the information to internal and external stakeholders, summarize these events. For activities of NMSS and FSME, NMED is an essential system used to collect information on such events.

The fuel cycle, materials, high-level waste repository, decommissioning, and spent fuel storage and transportation inspection programs are key elements in verifying the completeness and accuracy of licensee reports. The IMPEP also provides a mechanism to verify that Agreement States and NRC regions are consistently collecting and reporting such events as received from the licensees and are entering them in NMED.

The NRC has taken a number of steps to improve the timeliness and completeness of materials event data. These steps include assessment of the NMED data during monthly staff reviews; emphasis and analysis during the IMPEP reviews; NMED training in headquarters, the regions, and Agreement States; and discussions at all Agreement State and CRCPD meetings.

Validation:

There is a logical basis for using events involving radiation exposures to the public and occupational workers from radioactive material that exceed AO Criterion I.A as a performance measure for ensuring the protection of public health and safety. An event is considered an AO if it is determined to be significant from the standpoint of public health or safety. The NRC's regulatory process—including licensing, inspection, guidance, regulations, and enforcement activities—is designed to mitigate the likelihood of an event that would exceed AO Criterion I.A.

Events of this magnitude are rare. In the unlikely event that an AO occurs, NRC or Agreement State technical specialists, with input from expert consultants as necessary, will confirm whether the criteria were met.

The NRC believes that the data collected to meet this performance measure are free from bias. NMSS and FSME do not use statistical sampling of data to determine results. Rather, they review all event data to determine whether the performance measure has been met.

Two important data limitations in determining this performance measure are the delay time for receiving information and/or the failure of the NRC to become aware of an event that causes significant radiation exposures to the public or occupational workers. Although NMSS and FSME procedures and NRC regulations associated with event reporting include specific requirements for timely notifications, a lag time separates the occurrence of an event and the known consequences of that event.

The NRC believes that the probability of being unaware of an event that causes significant radiation exposures to the public or occupational workers is very small. Periodic licensee inspections and regulatory reporting requirements are sufficient to ensure that an event of this magnitude would become known.

If such an event occurs, it would result in a prompt and thorough investigation of the event, its consequences, its root causes, and the necessary actions by the licensee and the NRC to mitigate the situation and prevent recurrence. In addition to these immediate actions, the NRC holds periodic meetings where staff and management validate the occurrence of these events.

- number of radiological releases to the environment that exceed applicable regulatory limits, with a materials safety target of less than or equal to 5 and a waste safety target of 0

Verification:

This performance measure is defined as any release to the environment from fuel cycle, materials, high-level waste repository, decommissioning, and spent fuel storage and transportation activities that exceeds applicable regulations, as defined in 10 CFR 20.2203(a)(3). A 30-day written report is required regarding such releases. The nuclear materials safety performance measure target is less than or equal to five releases a year that meet this reporting criteria. The nuclear waste safety target is no releases that meet this reporting criteria.

If an event meeting this threshold occurs, it would be reported to the NRC and/or Agreement States through a number of sources, but primarily through required licensee notifications. Event notifications and preliminary notifications, which are used to widely disseminate the information to internal and external stakeholders, summarize these events. For activities of NMSS and FSME, NMED is an essential system used to collect information on such events.

The fuel cycle, materials, high-level waste repository, decommissioning, and spent fuel storage and transportation inspection programs are key elements in verifying the completeness and accuracy of licensee reports. The IMPEP also provides a mechanism to verify that Agreement States and NRC regions are consistently collecting and reporting such events as received from the licensees and entering them in NMED.

The NRC has taken a number of steps to improve the timeliness and completeness of materials event data. These steps include assessment of NMED data during monthly staff reviews; emphasis and analysis during the IMPEP reviews; NMED training in headquarters, the regions, and Agreement States; and discussions at all Agreement State and CRCPD meetings.

Validation:

The regulations in 10 CFR Part 20 provide standards for protection against radiation. There is a logical basis for tracking releases subject to the 30-day reporting requirement under 10 CFR 20.2203(a)(3)(ii) as a performance measure for ensuring the protection of the environment. The NRC's regulatory process—including licensing, inspection, guidance, regulations, and enforcement activities is sufficient to ensure that releases of radioactive materials that exceed regulatory limits are infrequent.

In the unlikely event that a release to the environment exceeds regulatory limits, the NRC or Agreement State technical specialists, with input from expert consultants as necessary, will confirm whether the criteria were met.

The NRC believes that the data collected to meet this performance measure are free from bias. NMSS and FSME do not look at statistical sampling of data to determine results. Rather, they review all event data to determine whether the performance measure has been met.

Two important data limitations in determining this performance measure are the delay time for receiving information and/or the failure of the NRC to become aware of an event that causes environmental impacts. Although NMSS and FSME procedures and NRC regulations associated with event reporting include specific requirements for timely notifications, a lag time separates the occurrence of an event and the known consequences of that event.

The NRC believes that the probability of being unaware of an event that causes a radiological release to the environment that exceeds applicable regulations is very small. Periodic licensee inspections and regulatory reporting requirements are sufficient to ensure that an event of this magnitude would become known.

If such an event occurs, it would result in a prompt and thorough investigation of the event, its consequences, its root causes, and the necessary actions by the licensee and the NRC to mitigate the situation and prevent recurrence. In addition to these immediate actions, the NRC holds periodic meetings where staff and management validate the occurrence of these events.

GOAL 2   SECURITY

Ensure the secure use and management of radioactive materials.

Strategic Outcome:

- No instances where licensed radioactive materials are used domestically in a manner hostile to the security of the United States

Performance Measure:

- Unrecovered losses or thefts of risk-significant[6] radioactive sources is 0.

Under FY 2007 AO Criterion I.C.1, the agency counts any unrecovered lost, stolen, or abandoned sources that exceed the values listed in Appendix P, "Category 1 and 2 Radioactive Material," to 10 CFR Part 110, "Export and Import of Nuclear Equipment and Material." Excluded from reporting under this criterion are those events involving sources that are lost, stolen, or abandoned under certain conditions, specifically (1) sources abandoned in accordance with the requirements of 10 CFR 39.77(c), (2) sealed sources contained in labeled, rugged source housings, (3) recovered sources with sufficient indication that doses in excess of the reporting thresholds specified in AO Criteria I.A.1 and I.A.2 did not occur during the time the source was missing, (4) unrecoverable sources lost under such conditions that doses in excess of the reporting thresholds specified in AO Criteria I.A.1 and I.A.2 were not known to have occurred, and (5) other sources that are lost or abandoned and declared unrecoverable; for which the agency has determined that the risk-significance of the source is low based on the location (e.g., water depth) or physical characteristics (e.g., half life, housing) of the source and its surroundings; where all reasonable efforts have been made to recover the source; and where it has been determined that the source is not recoverable and would not be considered a realistic safety or security risk under this measure.

Verification:

Losses or thefts of radioactive material that are greater than or equal to 1000 times the quantity specified in Appendix C, "Quantities of Licensed Material Requiring Labeling," to 10 CFR Part 20 must be

---

[6]   "Risk-significant" is defined as any unrecovered lost or abandoned sources that exceed the values listed in "Appendix P to 10 CFR Part 110–High Risk Radioactive Material, Category 2." Excluded from reporting under this criterion are those events involving sources that are lost or abandoned under the following conditions: (1) sources abandoned in accordance with the requirements of 10 CFR 39.77(c); (2) recovered sources with sufficient indication that doses in excess of the reporting thresholds specified in AO Criteria I.A.1 and I.A.2 did not occur during the time the source was missing; (3) unrecoverable sources lost under such conditions that doses in excess of the reporting thresholds specified in AO Criteria I.A.1 and I.A.2 were not known to have occurred; (4) other sources that are lost or abandoned and declared unrecoverable; (5) for which the agency has made a determination that the risk-significance of the source is low based upon the location (e.g., water depth) or physical characteristics (e.g., half life, housing) of the source and its surroundings; (6) where all reasonable efforts have been made to recover the source; and (7) it has been determined that the source is not recoverable and will not be considered a realistic safety or security risk under this measure.

reported (per 10 CFR 20.2201(a)) by telephone to the NRC Headquarters Operations Center or Agreement State immediately (interpreted as within 4 hours) if the licensee believes that an exposure could result to persons in unrestricted areas. If an event meeting the thresholds described above occurs, it would be reported through a number of sources, but primarily through this required licensee notification. Events that are publicly available are then entered and tracked in NMED, which is an essential system used to collect and store information on such events. Separate methods are used to track events that are not publicly available. Additionally, licensees must meet the reporting and accounting requirements in 10 CFR Part 73, "Physical Protection of Plants and Materials," and 10 CFR Part 74, "Material Control and Accounting of Special Nuclear Material."

The NRC's inspection programs are key elements in verifying the completeness and accuracy of licensee reports. The IMPEP also provides a mechanism to verify that Agreement States and NRC regions are consistently collecting and reporting such events as received from the licensees and are entering these events in NMED. In some cases, upon receiving a report, the NRC or Agreement State initiates an independent investigation that verifies the reliability of the reported information. When performed, these investigations enable the NRC or Agreement State to verify the accuracy of the reported data.

The regulation in 10 CFR 20.2201(b) requires a 30-day written report for lost or stolen sources that are greater than or equal to 10 times the quantity specified in Appendix C to 10 CFR Part 20 if the source is still missing at that time. In addition, 10 CFR 20.2201(d) requires an additional written report within 30 days of a licensee learning any additional substantive information. The NRC interprets this requirement as including reporting recovery of sources.

The NRC issued guidance in the form of a regulatory information summary (RIS 2005-21) to clarify the current 10 CFR 20.2201(d) requirement for reporting

recovery of a risk-significant source. FSME will ask the Agreement States to send copies of the RIS (or equivalent document) to their licensees. The NRC issued the National Source Tracking System final rule in November 2006. Implementation of this system will create and maintain an inventory of risk-significant sources. This rulemaking codifies and clarifies reporting requirements for risk-significant sources (including reporting timeframes) by adding specific requirements to 10 CFR 20.2201, "Reports of Theft or Loss of Licensed Material," for risk-significant sources, including a requirement for licensees to report the recovery of a risk-significant source within 30 days of recovery. In conjunction with this rulemaking, FSME will modify its Procedure SA-300 to specifically require Agreement States to report the recovery of a risk-significant source immediately to the NRC Headquarters Operations Center when notified by a licensee.

Validation:

Events collected under this performance measure are actual losses, thefts, or diversions of materials described above. Such events could compromise public health and safety, the environment, and the common defense and security. Events of this magnitude are expected to be rare. The information reported under 10 CFR Part 73 and 10 CFR Part 74 is required so that the NRC is aware of events that could endanger public health and safety or national security. Any failures at the level of the strategic plan would result in immediate investigation and follow-up.

If an event subject to the reporting requirements described above occurs, it would result in a prompt and thorough investigation of the event, its consequences, its root causes, and the necessary actions by the licensee, the NRC, and/or an Agreement State to mitigate the situation and prevent recurrence.

- Number of substantiated[7] cases of actual theft or diversion of licensed risk-significant radioactive

---

[7] "Substantiated" means a situation where an indication of loss, theft or unlawful diversion such as: an allegation of diversion, report of lost or stolen material, statistical processing difference, or other indication of loss of material control or accountability cannot be refuted following an investigation; and requires further action on the part of the agency or other proper authorities.

sources or a formula quantity[8] of special nuclear material or act that results in radiological sabotage is 0[9, 10]

Verification:

Substantiated means a situation where an indication of loss, theft or unlawful diversion such as: an allegation of diversion, report of lost or stolen material, statistical processing difference, or other indication of loss of material control or accountability cannot be refuted following an investigation; and requires further action on the part of the agency or other proper authorities. Licensees are required to call the NRC to report any breaches of security or other event that may potentially lead to theft or diversion of material or sabotage at a nuclear facility within 1 hour of its occurrence. The NRC's safeguards requirements are described in Section 73.71 of 10 CFR Part 73, "Physical Protection of Plants and Materials," and Appendix G to 10 CFR Part 73, "Reportable Safeguards Events," and in 10 CFR Part 74.11. The Information Assessment Team comprised of NRC Headquarters and Regional staff would conduct an immediate assessment for any significant events to determine what further actions are needed, including coordination with the intelligence community and law enforcement. The licensee is also required to file a written report within 30 days of the incident to describe the incident and the steps that the licensee took to protect the nuclear facility. This information would enable the NRC to adequately assess whether radiological sabotage has occurred. Any strategic plan failure results in immediate investigation and follow-up.

Validation:

Events that are required to be reported are those that endanger nuclear reactor facilities by deliberate acts of theft or diversion of material or sabotage directed against those facilities. Events of this type are extremely rare. If such an event occurred, it would result in a prompt and thorough investigation of the event, its consequences, its root causes, and the necessary actions by the licensee and/or NRC to mitigate the situation and prevent recurrence. The investigation ensures the validity of the information and assesses the significance of the event.

Verification:

In FY 2007 AO Criterion I.C.2, "substantiated" means a situation that requires additional action by the agency or other proper authorities because of an indication of loss, theft, or unlawful diversion—such as an allegation of diversion, report of lost or stolen material, statistical processing difference, or other indication of loss of material control or accountability—that cannot be refuted following an investigation. A formula quantity of special nuclear material is defined in 10 CFR 70.4, "Definitions." Radiological sabotage is defined in 10 CFR 73.2, "Definitions." Licensees subject to the requirements of 10 CFR Part 73 must call the NRC within 1 hour of an occurrence, to report any breaches of security or other event that may potentially lead to theft or diversion of material or to sabotage at a nuclear facility. The NRC's safeguards requirements are described in 10 CFR 73.71, "Reporting of Safeguards Events"; Appendix G, "Reportable Safeguards Events," to 10 CFR Part 73; and 10 CFR 74.11, "Reports of Loss or Theft or Attempted Theft or Unauthorized Production of Special Nuclear Material." The information assessment team composed of NRC Headquarters and regional staff members would conduct an immediate assessment for any significant events to determine any further actions that are needed, including coordination with the intelligence community and law enforcement. In accordance with 10 CFR 73.71(d), the licensee must also file a written report within 60 days of the incident describing the event and the steps that the licensee took to protect the nuclear facility. This information will enable the NRC to adequately assess whether radiological sabotage has occurred.

---

[8]   A formula quantity of special nuclear material is defined in 10 CFR 70.4.

[9]   "Radiological sabotage" is defined in 10 CFR 73.2.

[10]   Security goal performance measures 2, 3, and 4 together encompass the discontinued performance measure "Number of security events and incidents that exceed the Abnormal Occurrence Criterion I.C 2-4" to provide greater clarity and detail.

Validation:

Events subject to reporting requirements are those that endanger the public health and safety and the environment through deliberate acts of theft or diversion of material or through sabotage directed against the nuclear facilities that the agency licenses. Events of this type are extremely rare. If such an event occurs, it would result in a prompt and thorough investigation of the event, its consequences, its root causes, and the necessary actions by the licensee and/or the NRC to mitigate the situation and prevent recurrence. The investigation ensures the validity of the information and assesses the significance of the event.

- Number of substantiated losses of a formula quantity of special nuclear material or substantiated inventory discrepancies of a formula quantity of special nuclear material that are judged to be significant relative to normally expected performance or regulatory limits and that are judged to be caused by theft or diversion or substantial breakdown of the accountability system is 0.

Verification:

Licensees must record events associated with FY 2007 AO Criterion I.C.3 within 24 hours of the identified event in a safeguards log maintained by the licensee. The licensee must retain the log as a record for 3 years after the last entry is made or until termination of the license. The NRC relies on its safeguards inspection program to ensure the reliability of recorded data. The NRC makes a determination of whether a substantiated breakdown has resulted in a vulnerability to radiological sabotage, theft, diversion, or unauthorized enrichment of special nuclear material. When making substantiated breakdown determinations, the NRC evaluates the materials event data to ensure that licensees are reporting and collecting the proper event data.

Validation:

"Substantiated" means a situation that requires additional action by the agency or other proper authorities because of an indication of loss, theft, or unlawful diversion—such as an allegation of diversion, report of lost or stolen material, statistical processing difference, other system breakdown closely related to the material control and accounting program (such as an item control system associated with the licensee's facility information technology system), or other indication of loss of material control or accountability—that cannot be refuted following an investigation. A formula quantity of special nuclear material is defined in 10 CFR 70.4. Events collected under this performance measure may indicate a vulnerability to radiological sabotage, theft, diversion, or loss of special nuclear materials. Such events could compromise public health and safety, the environment, and the common defense and security. The NRC relies on its safeguards inspection program to help validate the reliability of recorded data and determine whether a breakdown of a physical protection or material control and accounting system has actually resulted in a vulnerability.

- Number of substantial breakdowns[11] of physical security or material control (i.e., access control containment or accountability systems) that significantly weaken the protection against theft, diversion, or sabotage is 0.

Verification:

For FY 2007 AO Criterion I.C.4, a "substantial breakdown" is defined as a red finding in the security oversight program or significant performance problems and/or operational events resulting in a determination of overall unacceptable performance or in a shutdown condition (inimical to the effective functioning of the Nation's critical infrastructure). Radiological sabotage is defined in 10 CFR 73.2. Licensees are required to report to the NRC,

---

[11] "Substantial breakdown" is defined as a red finding in the security inspection program, or any plant or facility determined to have overall unacceptable performance, or in a shutdown condition (inimical to the effective functioning of the nation's critical infrastructure) as a result of significant performance problems and/or operational events.

immediately after the occurrence becomes known, any known breakdowns of physical security, based on the requirements in 10 CFR 73.71 and Appendix G to 10 CFR Part 73. If a licensee reports such an event, the headquarters operations officer prepares an official record of the initial event report. The NRC begins responding to such an event immediately upon notification, with the activation of its information assessment team. A licensee must follow its initial telephone notification with a written report submitted to the NRC within 30 days.

The licensee records breakdowns of physical protection resulting in a vulnerability to radiological sabotage, theft, diversion, or loss of special nuclear materials or radioactive waste within 24 hours in a safeguards log maintained by the licensee. The licensee must retain the log as a record for 3 years after the last entry is made or until termination of the license. Licensees subject to 10 CFR Part 73 must also meet the reporting requirements detailed in 10 CFR 73.71. The NRC evaluates all of the reported events based on the criteria in 10 CFR 73.71 and Appendix G to 10 CFR Part 73. The NRC also maintains and relies on its safeguards inspection program to ensure the reliability of recorded and reported data.

Validation:

Events assessed under this performance measure are those that threaten nuclear activities by deliberate acts, such as radiological sabotage, directed against facilities. If a licensee reports such an event, the information assessment team evaluates and validates the initial report and determines any further actions that may be necessary. Tracking breakdowns of physical security indicates whether the licensee is taking the necessary security precautions to protect the public, given the potential consequences of a nuclear accident attributable to sabotage or the inappropriate use of nuclear material either in this country or abroad.

Events collected under this performance measure may indicate a vulnerability to radiological sabotage,

theft, diversion, or loss of special nuclear materials or radioactive waste. Such events could compromise public health and safety, the environment, and the common defense and security. The NRC relies on its safeguards inspection program to help validate the reliability of recorded data and determine whether a breakdown of a physical protection or material control and accounting system has actually resulted in a vulnerability.

- Number of significant unauthorized disclosures (loss, theft, and/or deliberate acts) of classified and/or safeguards information is 0.[12]

Verification:

With regard to FY 2007 AO Criterion I.C.5, any alleged or suspected violations by NRC licensees of the Atomic Energy Act, Espionage Act, or other Federal statutes related to classified or safeguards information must be reported to the NRC under the requirements of 10 CFR 95.57(a) (for classified information), 10 CFR Part 73 (for safeguards information), and NRC orders (for safeguards information subject to modified handling requirements). However, for performance reporting, the NRC would only count those disclosures or compromises that actually cause damage to the national security or to public health and safety. Such events would be reported to the cognizant security agency (i.e., the security agency with jurisdiction) and the regional administrator of the appropriate NRC regional office, as listed in Appendix A, "U.S. Nuclear Regulatory Commission Offices and Classified Mailing Addresses," to 10 CFR Part 73. The regional administrator would then contact the Division of Security Operations at NRC Headquarters, which would assess the violation and notify other NRC offices and other Government agencies, as appropriate. A determination would be made as to whether the compromise damaged the national security or public health and safety. Any unauthorized disclosures or compromises of classified or safeguards information that damage the national security or public health and safety would result

---

[12] "Significant unauthorized disclosure" is defined as a disclosure that harms national security or public health and safety.

in immediate investigation and follow-up by the NRC. In addition, NRC inspections will verify that licensees' routine handling of classified and safeguards information (including safeguards information subject to modified handling requirements) conforms to established security information management requirements.

Any alleged or suspected violations of this performance measure by NRC employees, contractors, or other personnel would be reported in accordance with NRC procedures to the Director of Division of Facilities and Security at NRC Headquarters. The NRC maintains a strong system of controls over national security and safeguards information, including (1) annual required training for all employees, (2) safe and secure document storage, and (3) physical access control in the form of guards and badged access.

Validation:

Events collected under this performance measure are unauthorized disclosures of classified or safeguards information that damage the national security or public health and safety. Events of this magnitude are not expected and would be rare. If such an event occurs, it would result in a prompt and thorough investigation, including consequences, root causes, and necessary actions by the licensees and the NRC to mitigate the consequences and prevent recurrence. NRC investigation teams also validate the materials event data to ensure that licensees are reporting and collecting the proper event data.

AGREEMENT STATES (AS OF AUGUST 2007)

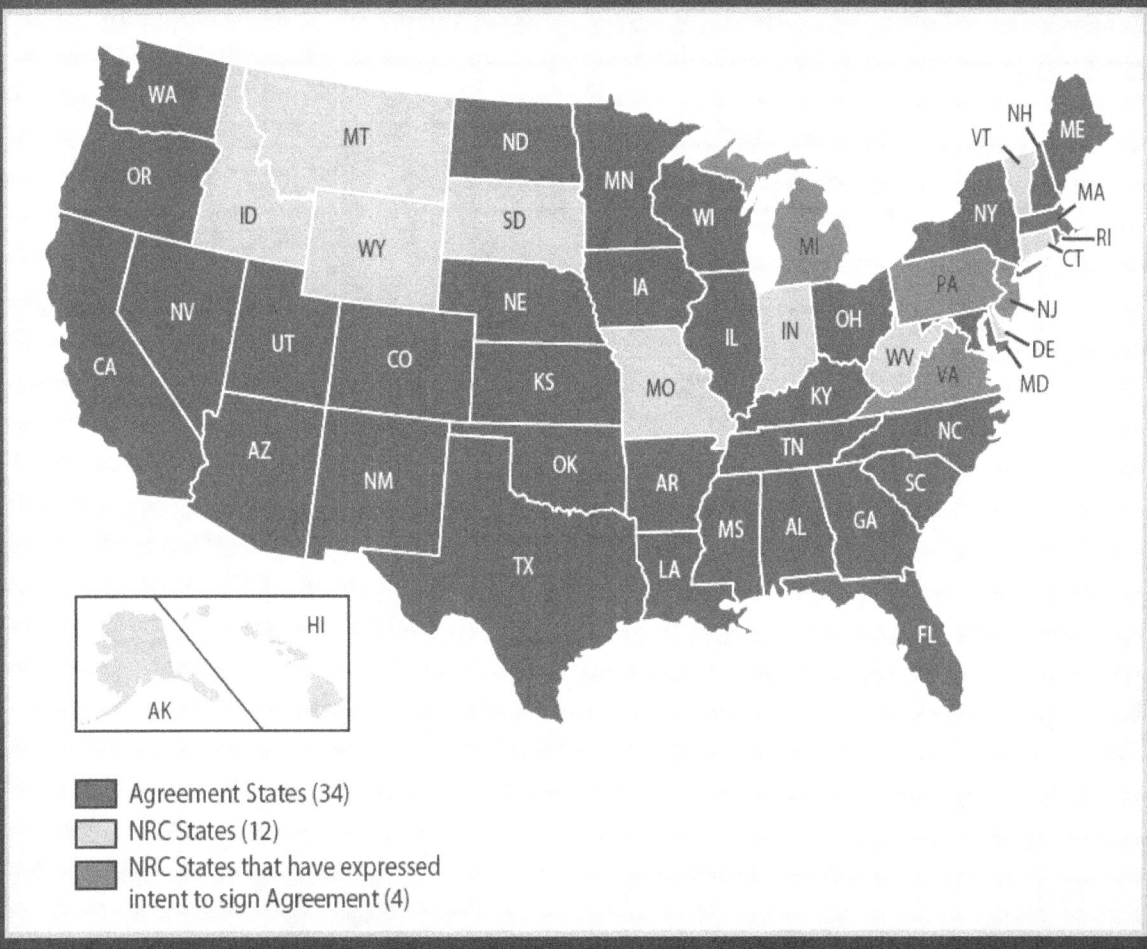

Agreement States (34)
NRC States (12)
NRC States that have expressed
intent to sign Agreement (4)

NRC ORGANIZATION CHART (AS OF AUGUST 2007)

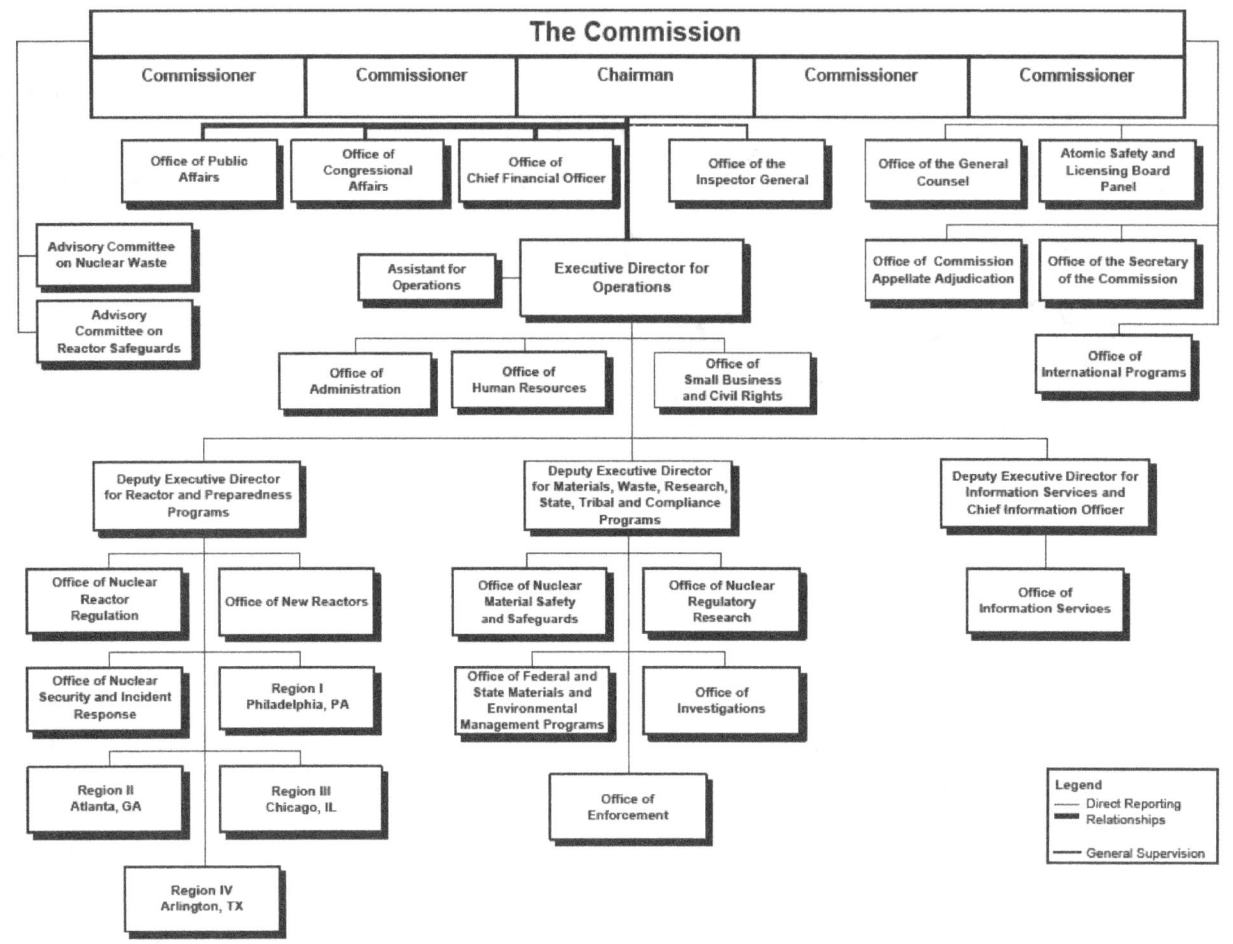

GLOSSARY OF ACRONYMS

ADAMS	Agencywide Documents Access and Management System
AICPA	American Institute of Certified Public Accountants
AO	abnormal occurrence
ASP	accident sequence precursor
BCP	Business Continuity Plan
CCR	Central Contractor Registration
CE	Combustion Engineering Owner's Group
CEAR	Certificate of Excellence in Accountability Reporting
CFO	Chief Financial Officer
CFO Act	Chief Financial Officer Act of 1990
CFR	United States *Code of Federal Regulations*
CIO	Chief Information Officer
CIOC	CIO Council
COLs	Combined Operating Licenses
CPIC	Capital Planning Investment Control
CSRS	Civil Service Retirement System
CY	calendar year
DHS	U.S. Department of Homeland Security
DOE	U.S. Department of Energy
DOI	U.S. Department of Interior
DOL	U.S. Department of Labor
EC	Executive Council
ECIC	Executive Committee on Internal Control

EDO	Executive Director for Operations
EFT	electronic funds transfer
e-gov	electronic Government
EO	Executive Order
EPA	Environmental Protection Agency
E-QIP	Electronic Questionnaires for Investigations Processing
ESP	Early Site Permits
FACTS I	Federal Agencies' Centralized Trial Balance System
FAR	Federal Acquisition Regulation
FECA	Federal Employees Compensation Act
FEMA	Federal Emergency Management Agency
FERS	Federal Employees Retirement System
FFMIA	Federal Financial Management Improvement Act
FFS	Federal Financial System
FICA	Federal Insurance Contribution Act
FISMA	Federal Information Security Management Act
FMFIA	Federal Managers' Financial Integrity Act of 1982
FOIA	Freedom of Information Act
FPPS	Federal Personnel and Payroll System
FSIO	Financial System Integration Office
FSME	Office of Federal and State Materials and Environmental Management Programs
FTE	Full-Time Equivalent
FY	fiscal year

GAO	U.S. Government Accountability Office	IT	information technology
GFE	Generic Fundamentals Examination	JFMIP	Joint Financial Management Information Program
GFRS	Governmentwide Financial Reporting System	LMS	Learning Management System
GLTS	General License Tracking System	LSN	Licensing Support Network
GPEA	Government Paperwork Elimination Act	MC&A	material control and accounting
GPRA	Government Performance and Results Act	MD	Management Directive
		MOX	mixed-oxide fuel
GSA	General Services Administration	MWe	Megawatts electric
GSI	General Safety Issue	NARA	U.S. National Archive and Records Administration
HHS	Health and Human Services	NBC	National Business Center
HLW	High-Level Waste	NFPA	National Fire Protection Association
HSPD	Homeland Security Presidential Directive	NIST	U.S. National Institute of Standards and Technology
HSPD-12	Homeland Security Presidential Directive 12	NMED	Nuclear Materials Event Database
IAEA	International Atomic Energy Agency	NMMSS	Nuclear Materials Management and Safeguards System
IG	Inspector General	NMSS	Office of Nuclear Material Safety and Safeguards
IMPEP	Integrated Materials Performance Evaluation Program	NRC	U.S. Nuclear Regulatory Commission
Improvement Act	Federal Financial Management Improvement Act of 1996	NRR	Office of Nuclear Reactor Regulation
Integrity Act	Federal Managers' Financial Integrity Act of 1982	NRO	Office of New Reactors
		NSIR	Office of Nuclear Security and Incident and Response
IOAA	Independent Offices Appropriation Act	NSTS	National Source Tracking System
IPAC	Intragovernment Payment and Collection	NUREG	Nuclear Regulatory Commission Regulation
IPSS	Integrated Personnel Security System	NWF	Nuclear Waste Fund
IRM	incident response manual	OBRA-90	Omnibus Budget Reconciliation Act of 1990
ISA	integrated safety analysis		

OCFO	Office of the Chief Financial Officer		RLO	records liaison officer
OEDO	Office of the Executive Director for Operations		RMG	records management guideline
			ROETF	Reactor Operating Experience Task Force
OIG	Office of the Inspector General			
OIS	Office of Information Services		ROP	Reactor Oversight Process
OMB	U.S. Office of Management and Budget		RTM	response technical manual
			SAT	Senior Assessment Team
OPM	U.S. Office of Personnel Management		SBR	Statement of Budgetary Resources
OSART	Operational Safety Review Team		SDLCM	System Development Life-Cycle Management
OUO	Official Use Only			
PAR	Performance and Accountability Report		SDLCMM	System Development Life-Cycle Management Methodology
PART	Program Assessment Rating Tool		SDP	Significance Determination Process
PBPM	planning, budgeting, and performance management		SECY	Office of the Secretary of the Commission
PC	Personal Computers		SFFAS	Statements of Federal Financial Accounting Standards
PII	personal identifiable information			
PL	Public Law		SGI	Safeguards Information
PMM	Project Management Methodology		SITSO	Senior Information Technology Security Officer
POA&M	plan of action and milestones		SNM	special nuclear material
PRA	Probabilistic Risk Assessment		SUNSI	Sensitive Unclassified Non-Safeguards Information
PRB	Petition Review Board			
PWR	Pressurized Water Reactor		TAC	Technical Assignment Control
RASP	Risk Assessment Standardization Project		TI	temporary instruction
			TSP	Thrift Savings Plan
RES	Office of Nuclear Regulatory Research		TSTF	Technical Specification Task Force
RIRIP	Risk-Informed Regulation Implementation Plan		USAID	U.S. Agency for International Development

NRC FORM 335 (2-89) NRCM 1102, 3201, 3202	U.S. NUCLEAR REGULATORY COMMISSION **BIBLIOGRAPHIC DATA SHEET** *(See instructions on the reverse)*	1. REPORT NUMBER (Assigned by NRC, Add Vol., Supp., Rev., and Addendum Numbers, if any.) NUREG-1542, Vol. 13

2. TITLE AND SUBTITLE

U.S. Nuclear Regulatory Commission
Performance and Accountability Report
FY 2007

3.	DATE REPORT PUBLISHED	
	MONTH	YEAR
	November	2007

4. FIN OR GRANT NUMBER
n/a

5. AUTHOR(S)

Richard Rough, et. al

6. TYPE OF REPORT

Annual

7. PERIOD COVERED *(Inclusive Dates)*
FY 2007

8. PERFORMING ORGANIZATION - NAME AND ADDRESS *(If NRC, provide Division, Office or Region, U.S. Nuclear Regulatory Commission, and mailing address; if contractor, provide name and mailing address.)*

Resource Mamagememt amd Support Staff
Office of the Chief Financial Officer
U.S. Nuclear Regulatory Commission
Washington, DC 20555-0001

9. SPONSORING ORGANIZATION - NAME AND ADDRESS *(If NRC, type "Same as above"; if contractor, provide NRC Division, Office or Region, U.S. Nuclear Regulatory Commission, and mailing address.)*

Same as 8, above

10. SUPPLEMENTARY NOTES

11. ABSTRACT *(200 words or less)*

The FY 2007 Performance and Accountability Report provides performance results and audited financial statements that enable Congress, the President, and the public to assess the performance of the agency in achieving its mission and stewardship of its resources. The report contains a concise overview, management's discussion and analysis, as well as performance and financial sections. Additional details of performance results and program evaluations can be found in the appendices.

12. KEY WORDS/DESCRIPTORS *(List words or phrases that will assist researchers in locating the report.)*

Performance and Accountability Report
FY 2007
PAR

13. AVAILABILITY STATEMENT
unlimited

14. SECURITY CLASSIFICATION

(This Page)
unclassified

(This Report)
unclassified

15. NUMBER OF PAGES

16. PRICE

NRC FORM 335 (2-89)

www.ingramcontent.com/pod-product-compliance
Lightning Source LLC
Chambersburg PA
CBHW081448170526
45166CB00008B/2357